Letting God Create Your Day

Volume 3
Paul A. Bartz

The Scripture quotations in this publication are from **The Holy Bible, King James Version,** copyright © by The World Publishing Company.

Letting God Create Your Day: Volume 3
Fourth Edition
by Paul A. Bartz

Copyright © 1992, 2006, 2015   Creation Moments, Inc.

Creation Moments, Inc.
P.O. Box 839, Foley, Minnesota 56329
www.creationmoments.com
800-422-4253

ISBN 1-882510-21-6

All rights reserved. No portion of this book may be reproduced, stored in a retrieval system, or transmitted in any form or by any means, electronic, mechanical, photocopying, recording, or otherwise, without prior written permission from the publisher.

Printed in the United States of America

Printing and production costs for this book were underwritten by supporters of Creation Moments.

Foreword

Why on Earth would Christian radio stations broadcast a program about sea lions that are learning how to do scientific research? For that matter, why would a Christian ministry such as Creation Moments publish a book that offers information on wasp reproduction? An explanation of how to wire hornets to power a digital watch might be fun to hear about. The Australian frog that swallows her eggs and hatches them in her stomach might tickle our curiosity. But what do these subjects have to do with Christian ministry?

No one would question Christians broadcasting and publishing a program on the Dead Sea Scrolls. It makes obvious sense for us to discuss the Christmas star at Christmas time. The discovery of Caiaphas's body is clearly newsworthy. A case can even be made for a story about Babylon rent-a-wagon.

Why waste time with stink bugs, animals that can live without air, and chimpanzees' math abilities? (Yes, all of these and many more appear in this volume.)

The Earth is the Lord's and the fullness thereof. In plain, modern language, the Earth and everything in it is the Lord's. Everything – electric wasps, singing lakes, plants that can swim and see, and fungi that weigh more than whales – was born in God's mind. So was the playful kitten.

Each created thing shows us a little about God. The immeasurable energy of the countless stars of heaven clearly shows His power.

God has created creatures that reflect a bit of one or another of His characteristics. Our inventive God has made creatures that show remarkable inventiveness. Consider that God made the entire universe in only six days. And He has also created creatures that reflect His industriousness in their own lives. The playful kitten and frolicking bear cub show us that God does indeed value humor and fun.

It's not hard to see God's hand in a sunset. It's sometimes a bit harder to see His hand in some parts of our lives. However, I maintain that when we learn to see God's hand in everything in the creation, we can more clearly see His hand in everything that happens in our lives. The realization that the Earth and everything in it belong to the Lord is also the realization that our lives and everything that makes up our lives are likewise in the Lord's hands.

Paul A. Bartz
Author, Creation Moments™

Worms with Kneecaps

Exodus 31:17
"'It is a sign between me and the children of Israel for ever: for in six days the LORD made heaven and earth, and on the seventh day he rested, and was refreshed.'"

New fossil discoveries in China are being greeted by evolutionists as among the most spectacular of the century. The fossils, say evolutionists, represent some of the earliest multicelled creatures. Evolutionists are publicizing these fossils as evidence for evolution. It's not difficult, however, to see how these fossils support creation rather than evolution.

Evolutionists admit the fossils show that the first multicelled creatures have appeared suddenly. This confirms creationist claims that life appeared suddenly and without evolutionary ancestors.

These "first" multicelled creatures were complex and complete. There were no fossils with partially developed eyes or other organs. They include trilobites that had some of the most complex eye structures of any creature that ever lived. There are also fossils of shrimp-like creatures and creatures with hard shells. Some of the animals were over two feet long. Now that's a big jump when you remember that the previous step was a single-celled algae!

There are no evolutionary missing links. The evolutionists said that the creatures all belong to groups that still exist. This means that science has now documented the history of some kinds of creatures alive today right back to the very first record of multicelled life. In other words, these creatures never evolved! Also of interest is how evolutionists explain the size of this fossil deposit, which is miles across. One evolutionist said, "A violent storm probably stirred up the sea bottom and the mud settled over a large area..."

Prayer: Dear Father in heaven, I do not understand how You could have designed and so quickly created the millions of kinds of living things that You did. Help me to believe Your Word, even when my poor, sinful, created mind cannot understand how wonderful You are. In Jesus' Name. Amen.

Ref: Wilford, John Noble. "Fast Evolutionary Jump Led to Complex Life, Study Says." *Star Tribune*, Wednesday, April 24, 1991, p. 4A.

Why Gazelles Stott

2 Samuel 2:18
"And there were three sons of Zeruiah there, Joab and Abishai, and Asahel: and Asahel was as light of foot as a wild roe."

When a gazelle sees a creature that threatens, it springs straight up into the air. If you have a house cat, you've probably seen your pet do something similar when someone tries to sneak up on it. This strange behavior is called stotting.

Scientists are well aware that features found in nature have purpose. This is not to say, however, that all scientists necessarily believe in a Creator.

At first glance, it would seem that stotting is counterproductive. Time and energy that might be used in running away from a predator seem to be wasted in springing up, making the prey even more noticeable. Is stotting an animal's attempt to warn others of its kind about danger? Is it an attempt to get a better view of the danger? Or, could stotting be an attempt to confuse the predator? After watching 250 incidences of gazelle stotting, naturalists believe they have come up with a solution. They believe that stotting is a creature's way of letting a would-be predator know that he's been seen. The message is, "don't try to sneak up on me, because I know you're there." Researchers noted that stotting does seem to discourage cheetahs from continuing to stalk a gazelle.

Even though most scientists don't believe in the Creator, they sometimes recognize that there is intelligent purpose and design in the creation. How contradictory it is for those who would deny the Creator to then use what He has made to try to deny Him!

> **Prayer: Dear Lord, Your wisdom in protecting and caring for Your creation goes far beyond our puny human wisdom. I thank You for that. I ask that You would never let me forget that Your wisdom and love are both greater than I can comprehend. Amen.**

Ref: "Verily, Gazelles Leap After they Look." *Discover*, Sept. 1986, p. 13.

A Brilliant Escape

1 Corinthians 10:13
"There hath no temptation taken you but such as is common to man: but God is faithful, who will not suffer you to be tempted above that ye are able; but will with the temptation also make a way to escape, that ye may be able to bear it."

It is always as dark as night 2,000 feet beneath the ocean's surface. Yet, a rich variety of life thrives in the darkness. The strategies for life far beneath the waves are remarkable and unique.

Even in the deep ocean, predator searches for prey and prey tries to escape predator. The permanent darkness and buoyant water allow for unusually creative strategies in this rhythm of life. Some creatures try to hide in the darkness. Others use light to lure prey close to them. Some creatures even squirt luminescent clouds into the water to make predators think that they are where they're not.

Perhaps the most creative is a species of jellyfish. When a potential predator comes close to the jellyfish, the jellyfish turns off the lights in its bell-shaped body. Then, switching off the lights in its tentacles, the jellyfish scoots away as fast as it can. If the predator wasn't fooled and continues the chase, the jellyfish switches to plan B. It turns on the blue light in its body and the white lights in its tentacles. When the attacker is very close, the jellyfish turns off the light in its body and takes off, detaching its still-glowing tentacles. The tentacles continue to writhe in the water, distracting the predator.

According to evolution theory, jellyfish are among the most primitive multicelled animals. In this clever survival strategy we see that jellyfish are neither simple nor primitive. They are a true credit to the Creator's handiwork!

Prayer: Father, nothing You have made fails to declare Your glory. Help me do a better job of glorifying You so that more people may be invited to Your forgiving grace in Jesus Christ. In His Name. Amen.

Ref: *Discover*, February, 1986, p. 67.

Petrifying Ages

1 Corinthians 10:4
"And did all drink the same spiritual drink: for they drank of that spiritual Rock that followed them: and that Rock was Christ."

How long does it take to petrify wood? Scientists who believe in those millions and billions of years that evolutionists are always talking about have never tested the answer to this question. They simply assumed that it must take hundreds or thousands of years to petrify wood. It wasn't until the 1970s that scientists bothered to explore this question.

Bone, wood, and other once-living materials become petrified when silica replaces the original organic material. It has always been assumed that silica is not very active at temperatures below the boiling point of water. And for the same reason, silica rock formations were thought to take thousands of years to form.

In the early 1970s, a paper was published showing that silica crystals could be grown in backyard conditions in only three years! This means that wood or bone can turn into stone within a few years. It also means that rock formations once thought to take thousands of years to form could have been formed in a relatively short time. All that's needed are silica and water.

Here is an example of how belief in evolution has retarded science. If no one had ever questioned the long ages claimed by evolution, we would still be ignorant about how silica works in nature around us. This research also helps uphold the Bible's claims that the Earth and life have only been around for thousands, not millions, of years. The truth of God's Word remains unchallenged!

Prayer: Dear Lord Jesus Christ, I thank You that Your Word to us can be trusted. Man's word can mislead, for men are easily misled. Help me to apply Your Word to my life so that I may be led by You and not by human error. Amen.

Ref: Chittick, Donald E. The Controversy. *Multnomah Press*, 1984, pp. 240-241.

What's a Siphonophore?

Numbers 23:19
"'God is not a man, that he should lie; neither the son of man, that he should repent: hath he said, and shall he not do it? or hath he spoken, and shall he not make it good?'"

It can be forty feet long. It can have many mouths and just as many stomachs. It swims along in the darkness more than 1,500 feet beneath the sea, reaching out for food with its lethal tentacles.

This is not a creature invented for a new horror film. It's a siphonophore. As one scientist described it, "A siphonophore is this big thing that goes around sucking up food." In reality, a siphonophore is a collection of jellyfish that join into a huge colony. When they link up, some of the jellyfish act as mouths, others act as stomachs. Other jellyfish in the colony take care of the swimming while yet others cast out their tentacles to gather food. When joined, they act as a single, huge creature.

The siphonophore, however, is not the terror of the deep to all creatures. When the siphonophore's tentacles are extended for feeding, lantern fish and smelt gather. The smelt actually snuggle harmlessly among the tentacles. These fish are waiting for the scraps that result from the siphonophore's feeding. During its feeding, the siphonophore also provides refuge for fish hiding from predators.

God can make anything He wants. There are no limits to His imagination and power. Many people have said to us, doesn't an all-powerful, unlimited God mean that God could have created everything any way He wanted? Couldn't God have used evolution? The answer is, "Yes." But the God who created the universe itself is also the God who got it right the first time, every time. He said it was perfect. He did not do it by trial and error.

Prayer: I thank You, Lord, that Your Word is trustworthy and that You cannot lie or mislead me. Grant me Your Holy Spirit to teach me when I study Your Word so that I may not be misled by my own thoughts. Amen.

Ref: *Discover*, February, 1986.

Extinct Tree Is Thriving

Isaiah 55:8
"For my thoughts are not your thoughts, neither are your ways my ways, saith the LORD."

According to the claims of evolutionists, the dawn redwood trees lived from the time of the dinosaurs until about two million years ago. Then they became extinct. At the very same time this was the official scientific teaching, Chinese rice farmers were planting the tree because it was a good indicator of fertile rice fields.

Evolutionary scientists caught up with the real world in 1944, when they acknowledged the tree's existence. In 1947, Chinese botanists sent dawn redwood seeds to Harvard University. Today the dawn redwood is a popular landscape tree across America. This slender tree, with its gently drooping branches and bright green, feathery-looking needles, looks like a living Chinese painting. It grows up to four feet per year. Unlike most conifers, the dawn redwood drops its needles in the fall.

You might correctly ask, "Can't scientists be wrong sometimes?" Yes! Like the rest of us, scientists are only human. That's exactly the lesson. If scientists can be wrong about the present, they have no business telling us that the history recorded in the Bible cannot possibly be true. If a tree as important as the dawn redwood could escape their notice for so long, how can they tell us that life existed for millions of years before the first human beings?

Yes, scientists are human like the rest of us. That means that like the rest of us, their proper place is one of humility and thanksgiving before the Creator and Lord of the universe.

Prayer: Dear Father, I confess that it is not only those who would deny You who think too highly of their own knowledge and abilities. I have done the same. Forgive me for the sake of Your Son, Jesus Christ, and fill me with Your peace. Amen.

Ref: Wolf, Thomas H. "The Object at Hand." *Smithsonian*, Sept. 1990, pp. 26-28.

Pure Pain

Revelation 21:4
"And God shall wipe away all tears from their eyes; and there shall be no more death, neither sorrow, nor crying, neither shall there be any more pain: for the former things are passed away."

Despite modern medical advancements, pain management remains a difficult problem today. Man has used various forms of aspirin to manage pain for thousands of years. While more potent drugs can be used today, our primary approach to pain has changed little over the millennia. Recent advancements now promise to show us how to deal directly with many forms of pain.

Let's say that you accidentally get a paper cut. The pain produced by a paper cut is always out of proportion to the actual injury. But medical researchers tell us that the injured tissue creates a chemical called *bradykinin*. Bradykinin is the most potent pain-producing substance known. Bradykinin has a greater purpose than simply making your paper cut miserable. Bradykinin sets off a complex chain of chemical reactions that are important in healing your wound. In the process, bradykinin also links to nerve cells, causing them to send pain messages to your brain.

Scientists have begun to develop bradykinin blockers. These chemicals are designed to bind to the nerve cells in place of the bradykinin, preventing bradykinin from sending pain messages. As a result, pain is blocked at the source.

Pain is a reality in our world because of sin. It not only warns us of injury but it reminds us of the ultimate consequences of sin. But thanks be to God who sent His Son Jesus Christ to bear the pain of our disobedience on the cross so that we might have the forgiveness of sins!

Prayer: I thank You, dear heavenly Father, that You loved me so much that You sent Your beloved Son, Jesus Christ, to suffer the penalties of my sin. Help me to always seek my assurance of Your acceptance in what He has done for me. Amen.

Ref: McKean, Kevin. "Pain." Discover, October, 1986, pp. 82-92.

Doctor Frog

Ephesians 6:14-16
"Stand therefore, having your loins girt about with truth, and having on the breastplate of righteousness; And your feet shod with the preparation of the gospel of peace; Above all, taking the shield of faith, wherewith ye shall be able to quench all the fiery darts of the wicked."

The accidental discovery of a frog's secret may provide a giant leap for medical science.

For years, biochemist Michael Zasloff had been surgically removing the ovaries of African clawed frogs for research. After surgery, the frogs were returned to their tanks. The tanks offered conditions very different from a hospital recovery room. The water in the tanks teemed with bacteria, parasites, fungi, and viruses. Yet, the frogs' surgical wounds healed without infection almost every time. After five years, Zasloff finally realized that something unusual was going on.

Frogs, like humans, have an immune system. The normal immune system, however, does not work well at the low temperatures in which frogs are usually found. So Zasloff began studying the frogs' skin. After months of work, he finally discovered two previously unknown proteins in the frogs' skin that kill a wide range of bacteria and fungi. He learned that the frogs have a second, independent defense system to help them live in a potentially fatal environment.

Zasloff named the new proteins he discovered *magainins*. He chose the term from the Hebrew word *magain,* which means "shield." In the Old Testament, God is often called our "magain," our shield. Clearly His special love and care in helping His creatures in a dangerous environment extends even to frogs!

Prayer: My Lord and Savior Jesus Christ, You are my shield from sin, death, and the devil. Let me never doubt Your love and protection, even in the most dangerous of conditions. If You so care for frogs, I know you care even more for me! Amen.

Ref: Beard, Jonathan. "The Good and the Bad of the Ugly." *Discover*, Jan., 1988, p. 42.

Confused Birds

Romans 8:20-21
"For the creature was made subject to vanity, not willingly, but by reason of him who hath subjected the same in hope, Because the creature itself also shall be delivered from the bondage of corruption into the glorious liberty of the children of God."

We are all aware of migrating birds' amazing abilities to navigate thousands of miles to a specific forest or tree. However, do migrating birds ever make mistakes in navigation? The answer is yes. Those mistakes help us understand how birds navigate.

Birds that make mistakes in navigation are called *vagrants*. Many of the vagrant warblers found in California nest in Canada, east of the Canadian Rockies. In the fall, they fly south or southeast, either to the Atlantic coast or the Gulf of Mexico, and finally to their wintering grounds in Central or South America. For some unknown reason, about 150 of the birds usually end up in California.

To find an explanation, scientists studied the starting ranges of these birds in Canada. Their first discovery was that most of the vagrant birds were first-timers. But since compass headings for migration are genetically built into the birds, how could they get lost? The problem was that the birds confused their left and their right! If, for example, a species was programmed to fly 55 degrees to the *left* of south, the vagrants arrived in California by flying 55 degrees to the *right* of south.

Vagrant birds are one illustration of how man's sin has placed a burden of decay upon the entire creation. We can understand why all the creation eagerly awaits the return of Christ for its complete deliverance from the burden of sin.

Prayer: I know, Lord, that it is because of Your mercy that You tarry in Your return to us. You would have no one unnecessarily lost. Help me to do a better job of witnessing to others about forgiveness, even as I pray, come quickly, Lord. Amen.

Ref: Diamond, Jared. "The Case of the Vagrant Birds-or. Left Coast, Here we come." *Discover*, January, 1986, pp. 82-84.

Toxic Butterflies Fool Evolutionists

John 3:12
"'If I have told you earthly things, and ye believe not, how shall ye believe, if I tell you of heavenly things?'"

The monarch caterpillar feeds on milkweed. Milkweed manufactures a powerful toxin that can, in most cases, stop the heart of any creature who eats enough of it. The monarch caterpillar itself is unharmed by this poison. In fact, the caterpillar stores the poison in its body, and this poison remains even after the caterpillar has turned into a butterfly.

Evolutionary scientists thought that the viceroy evolved to mimic or look like the monarch to fool birds into thinking that it, too, was toxic. The evolutionary story was that viceroys must really be good to eat, since they evolved from the tasty admiral butterflies. This evolutionary thinking remained untested until a few years ago, basically because scientists often consider it improper to question evolutionary claims. But the tests using the wingless bodies of six different kinds of butterflies, including viceroys, proved that viceroys are indeed toxic to birds. Birds avoid the viceroy because it manufactures its own toxins. In fact, research has shown that the viceroy is, on the average, even more poisonous than the monarch!

Evolution is bad science. In this example we see how evolution led to generations of misunderstanding about how God protects the viceroy butterfly.

Prayer: Lord, where the truth of Your Word is rejected, ignorance about even Your creation soon follows. I pray that the darkness and ignorance caused by evolution would be reduced through the bold witness to the truth by Your people. Amen.

Ref: Walker, Tim. "Butterflies and Bad Taste." *Science News*, Vol. 139, p. 348.

Your Portable First Aid Kit

Nehemiah 9:6
"Thou, even thou, art LORD alone; thou hast made heaven, the heaven of heavens, with all their host, the earth, and all things that are therein, the seas, and all that is therein, and thou preservest them all; and the host of heaven worshipeth thee."

Ouch! You've just gotten a paper cut on your finger. What's the first thing you do? If you're like most people, you'll probably put your finger in your mouth. If you think about it, you probably have no idea why you put your finger into your mouth. Actually, when a dog licks its wounds or when you put your paper cut-ravaged finger in your mouth, you are beginning medical treatment.

Medical science is only just learning what God knew when He built this reaction into us. The saliva of mammals and human beings contains epidermal growth factor. Studies show that when epidermal growth factor is applied to wounds, healing takes place much faster. Epidermal growth factor increases the number of cells available to grow new skin over a wound. It also encourages capillaries to form near the wound to increase blood supply. Epidermal growth factor doubles the amount of new DNA at the wound site. Finally, it increases the amount of collagen in the wound to give the new tissue the strength to close up and remain closed.

Researchers have little idea how epidermal growth factor works. They are still researching this powerful medicine that's found in our saliva.

Only our wise and loving Creator could have provided us with this convenient and powerful medical treatment. Only He could have built into us the natural reaction of putting an injured finger into our mouth. Like so much else in the creation, this blessing is too well designed to be nothing more than a series of unrelated coincidences.

Prayer: Dear Father, our pagan world talks about chance and coincidence. But I learn in Your Word that there is no such thing. Cleanse my thinking and speech of this pagan influence so that I may better witness Your truth. In Jesus' Name. Amen.

Ref: "An Aid to Healing that Simply Can't be Licked." *Discover*, April, 1986, p. 10.

A Glowing Ballet

Ephesians 2:10
"For we are his workmanship, created in Christ Jesus unto good works, which God hath before ordained that we should walk in them."

Many creatures have complex and beautiful mating rituals. Few are more wondrous or dramatic than the firefleas' ritual.

The fireflea is a marine crustacean about the size of a sesame seed. They are found in the Caribbean. Firefleas, as the name implies, are capable of generating light. The light is used to signal danger and for their mating ritual. When attacked, firefleas generate a cloud of light in the water, helping some firefleas escape and drawing the attention of larger predators to the firefleas' attacker. There is enough light-generating chemical in one fireflea to give ten minutes of light bright enough to read by.

During the mating ritual, only the males emit light. But this is not simply random lighting. The males produce differing dot and dash patterns in the water as they try to gain a female's approval. Some species produce vertical lines of dots while others make horizontal lines. One species swims in packs, producing large dotted patterns by coordinating their release of light with each other.

The fireflea is a tiny and seemingly unimportant creature. The meaning of its complex light displays went unnoticed until just the last few years. Though unknown to man for thousands of years, the fireflea has continued to praise its Creator by doing what He created it to do. Are you doing what your Creator created you to do?

Prayer: I thank You, dear Father, that through the forgiveness of my sins in Jesus Christ, You have made me a new creature. Help me to know how to live the new life You have given me in Christ so that I may praise You with my actions. Amen.

Ref: "Romantic Lighting." *Discover*, Feb. 1988, p. 16.

Do Spiders Feel Pain?

1 Peter 3:18
"For Christ also hath once suffered for sins, the just for the unjust, that he might bring us to God, being put to death in the flesh, but quickened by the Spirit:"

If you are like most people, you may not care whether spiders feel pain. Because of the way spiders are designed, however, their ability to feel pain has implications for many other creatures.

Like crabs and lizards, spiders are able to drop a leg when attacked. Spiders, too, have enemies, among them the poisonous ambush bug. Scientists have always assumed that a spider would drop a leg bitten by the ambush bug to prevent the bug's venom from spreading to the rest of its body. But, when they watched an ambush bug attack a spider, it seemed that the spider reacted in pain to the bite.

This observation led scientists to test the spider's reaction to various parts of the ambush bug's venom. Four components of the venom are both pain-producing and poisonous. Two other components produce pain, but are harmless. Others cause no pain. Spiders detach a leg only when it is injected with the components that are both poisonous and painful. Scientists were surprised that spiders could feel pain. They were also surprised that the same substances that cause pain in humans also cause pain in spiders.

Science has, for too long, ignored the obvious fact that other living things can feel pain. Some have even taken this so far as to say that the unborn human child cannot feel pain when attacked during an abortion. The spider's proven ability to feel pain teaches all of us to be less unfeeling about the pain of others.

Prayer: Dear Lord, I thank You for bearing the pain of my sin on the cross. Help me to be more empathetic, loving, and sensitive to the pain of others, including animals. Help me to be more like You! Amen.

Ref: "The Sensitive Spider." *Science 83*, p. 6.

Bug Baits Bug

1 Timothy 6:9
"But they that will be rich fall into temptation and a snare, and into many foolish and hurtful lusts, which drown men in destruction and perdition."

All of us are familiar with using bait to lure prey. Both live and artificial bait are used to catch fish. Stores lure shoppers with sale prices. The use of both bait and camouflage is an even more sophisticated activity. The duck hunter not only hides in a blind disguised to look like the natural surroundings but also sets decoys around him to make it look as if the area is safe for ducks.

The juvenile assassin bug is just as sophisticated as the duck hunter. Assassin bugs are very fond of termites. A juvenile assassin bug will glue bits of the termites' nest to its body until it is nearly invisible against the nest. The bug will then patiently wait for a termite to wander too close. The first termite caught is drained by the assassin bug and then its body is dangled in front of an opening in the nest.

When a nest mate dies, termites usually save the body as a protein source in their protein-poor diet. As a termite grabs for the body of its nest mate, the assassin bugs pulls the body back. The termite is lured further out of the nest until it is grabbed for lunch. The second termite is then used as bait. Scientists report that they once watched a single assassin bug dispose of 31 termites in three hours!

The juvenile assassin bug is every bit as clever about its hunting as the human duck hunter. In its own way, this bug proves that intelligence did not evolve; it is a gift of our all-wise Creator.

Prayer: Grant me, dear Father, a thirst for Your Word, so that I may know Your truth and wisdom. I ask that my immersion in Your Word may prepare me to resist and flee the lures of the devil so that I may serve You now and forever. In Jesus' Name. Amen.

Ref: "Bugs that Use Bait." *Science 83,* p. 6.

Joshua's Altar

Joshua 8:30
"Then Joshua built an altar to the LORD God of Israel in Mount Ebal,"

In Deuteronomy 27:2-8 we read of how Moses instructed Joshua to build an altar on Mt. Ebal after the Israelites had entered the Promised Land. There the people were to sacrifice and give thanks to the Lord.

Bible critics have generally been skeptical of the Bible's early history of Israel's conquest of Canaan. Over the years, however, a growing number of discoveries have supported the Bible's account of the earliest history of the Israelites. Perhaps one of the most dramatic recent discoveries is the altar Joshua apparently built on Mt. Ebal after Israel entered Canaan.

The ruins of the altar are laid out in a rectangle that measures 25 by 30 feet. The walls of the nine-foot-high structure of uncut stones and rocks are filled with ash, dirt, stones, and bones. Over 4,000 bones were found in the fill, all of them bones of the sacrificial animals called for in Moses' law. Evident, too, is the ramp up to the altar that was used by the priests, as stipulated by Exodus. One discovery at the site seems to tie everything together. It is an Egyptian scarab that was probably among the spoils the Bible says Israel took with them from Egypt.

Bible-believing Christians don't need this proof to accept the Bible. But it is exciting to identify and learn more about such a historic place as the Mt. Ebal altar.

Prayer: Dear Father in heaven, I know that Your Word is trustworthy. But I thank You that You have provided this additional witness to the integrity of Your Word for those who doubt. Let them, too, come to know Your grace in Christ. Amen.

Ref: Machlin, Milt. "Joshua and the Archaeologist." *Reader's Digest,* Sept., pp. 135-138.

What the Unborn Tells Mother

Jeremiah 1:5
"Before I formed thee in the belly I knew thee; and before thou camest forth out of the womb I sanctified thee, and I ordained thee a prophet unto the nations."

When talking about a baby's expected time of birth, we have all heard someone say, "The baby will come when it's ready." While that might not sound very scientific, new research shows that the statement is probably scientifically accurate. The infant, not the mother, seems to control the start of the birth process for both mother and infant.

Medical researchers are doing more research to confirm their theories. They believe that the brain of the unborn child monitors the infant's development so that it knows when the infant is ready for birth. As the time for birth approaches, a pea-sized part of the fetal brain signals the adrenal and pituitary glands that it is soon time to be born. These glands respond by producing two hormones. These hormones build up in the infant's blood. As they build up, they create changes in the mother's hormones that begin the process of giving birth.

A complete understanding of this amazing sequence may help medicine treat the problem of premature labor.

More important, we see how exquisitely God has designed the process of giving birth. How wise of God to design the process so that when the baby is ready to be born, the mother's *convenience* must give way to the baby's *need.* Nor does the baby's birth time depend on the mother's limited understanding of the developing infant. Perhaps these realities offer some lessons to an age that places so much less value on the unborn child than God clearly does.

Prayer: Lord, I pray for an end to the murder of the unborn through abortion. Help Your people to live their faith so clearly and invitingly that those who reject Your truth would be invited to the forgiveness of sins by the joy and excellence of our lives. Amen.

Ref: Fackelmann, K. "A. Fetus Tells Mother: It's Time for Labor." *Science News*, Vol. 140, p. 182.

Science Proves Teaching Abstinence Works!

Proverbs 22:6
"Train up a child in the way he should go: and when he is old, he will not depart from it."

Popular folk wisdom, heavily promoted by groups like Planned Parenthood, says that it is pointless to teach teenagers that premarital sex is wrong. They claim that young people will experiment with sex no matter what they are told. Now the first broad-based, long-range study of this folk wisdom has proven it to be a myth.

The five-year study, conducted by an Illinois group called Project Respect, involved 3,500 students at 26 schools. It concluded that middle school sex education that stresses sexual abstinence before marriage *does* reduce teen pregnancy. The textbook used in the course was *Sex Respect: The Option of True Sexual Freedom.* It teaches students that abstinence before marriage offers the healthiest approach to life, especially in view of sexually transmitted disease, emotional trauma and pregnancy.

The study found that even two years after the class, the pregnancy rate for girls who took the class was half the national average of ten percent. Also, fewer males were involved in causing pregnancies. There was a twenty percent increase in the number of students who agreed, two years after the course, that "sexual urges are always controllable." Other questions also revealed the development of healthier attitudes among teenagers about sexually transmitted diseases and marriage.

Parents and educators should be pleased to learn that education in decency, and healthy attitudes toward the gift of sex does work. Perhaps modern educators are finally learning a principle that has been taught by the Bible for thousands of years.

Prayer: For too long, dear Lord, too many of Your people have been intimidated by the false claims of those who reject the truth. Forgive us for being timid and make us bold to speak the truth so that Christian young people may not be taken captive by the devil. Amen.

Ref: "Study Show Teaching Abstinence Works." *Educator Reporter*, May 1991.

The Woodpecker's Pantry

1 Kings 17:14
"For thus saith the LORD God of Israel, The barrel of meal shall not waste, neither shall the cruse of oil fail, until the day that the LORD sendeth rain upon the earth."

Many kinds of birds collect and store food for later use. Now, scientists report that the female red-cockaded woodpecker collects, stores, and appropriately uses a dietary mineral supplement as well.

As is true for most birds, the female red-cockaded woodpecker needs additional dietary calcium at egg-laying time. Scientists tracking the female woodpeckers noticed that within a couple days of egg-laying time, the woodpeckers began gathering and hiding bone fragments found on the forest floor. Upon spying a bone fragment, the female would usually consume a few flakes of it on the spot and then carry the rest off to her "pantry" for later use. As she laid her eggs, she would frequently return to her pantry and eat more flakes from the bone. Once egg laying was completed, the female woodpeckers showed little interest in the bone pieces.

Scientists noted that this was the first time a bird had been found to collect and store dietary mineral supplements. They pointed out that by collecting the bone from the forest floor and storing it in trees near the nest, the female woodpeckers avoided the risk of attack by floor-dwelling predators.

A healthy life requires planning and maintaining health balances. For this reason the Creator's love for all He has made led Him to give even the animals a detailed knowledge of how to stay healthy.

Prayer: Father in heaven, help me to love all that You have made the way You do. I know that my love can never be perfect, like Yours. But I also know that my care of Your creation and concern for living things can improve. In Jesus' Name. Amen.

Ref: "Red-cockaded Woodpeckers Stash Bone." *Science News*, August 31, 1991, p. 143.

Will Mammoths Walk the Earth Again?

Genesis 1:24
"And God said, Let the earth bring forth the living creature after his kind, cattle, and creeping thing, and beast of the earth after his kind: and it was so."

Mammoths and mastodons were distant relatives to the elephant. Today, there is a great deal of debate over when they became extinct and why.

Mammoths and mastodons are less mysterious to us than dinosaurs because we can study their preserved bodies. Several of the creatures have even been found, complete with flesh, frozen in the ice in Siberia, North America, and northern Europe. Dogs have even found this flesh good to eat. Much of the ivory available today has been taken from the dead mammoths. Mammoths seem almost, but not quite, part of our world. That could change.

Soviet scientists removed living sperm from some mammoths that had fallen into an icy pond and been almost perfectly preserved since their deaths. The sperm was used to fertilize nine elephant eggs. The eggs were then re-implanted in the elephants. Only one of the nine survived until birth. The half mammoth-half elephant was sterile, small, and covered with shaggy yellow hair. Soviet scientists are continuing their research. They hope to find and fertilize mammoth eggs with mammoth sperm. Elephants would be used as surrogate mothers. If they are successful, mammoths might again walk our Earth!

The history of life on Earth shows that the further back in time we go, the more kinds of creatures existed. This is the opposite of what we would expect if evolution were true. But, it is exactly what we would expect based on the Bible's account of creation.

Prayer: I thank You, Lord, for the rich variety of life that surrounds us. I pray that we might do a better job of using extinct creatures to glorify You rather than using them to call Your work of creation into question. Amen.

Ref: "What is a Fossil?" *The Rich, Rich Desert*. Student Handout.

Selfishness Loses Out

Philippians 2:3
"Let nothing be done through strife or vainglory; but in lowliness of mind let each esteem other better than themselves."

There is no room in the harsh realities of evolution for selflessness. For example, evolution says that if I help you, I do so because I'm going to get something for myself. Likewise, what appears to be friendly behavior among animals is said to be only an evolutionary adaptation designed to preserve an animal's genetic code. Now, new research is leading some to question this harsh materialism.

Black Mediterranean ants live in colonies ruled by a queen. Their young are all genetically related to the rest of the colony. As the young are cared for, they are carried about the colony and licked. Researchers wanted to find out whether unrelated larvae from other colonies would receive as much care as larvae raised by their kin.

The scientists transferred eggs from one colony to another, where non-kin cared for them until they reached the larval stage. They were then returned to their own nest. When these ants became adults, researchers gave them the choice of caring for the young of their kin or the young from the colony where they were nursed. These ants gave more care to their non-kin than to their own family. Researchers believe that this is because they remembered the scents of those who nursed them.

The idea that all relationships and altruism are a result of selfishness is a cold, hard way to look at life. It also ignores the features God has built into the creation so that life would not be a desolation of cold, hard selfishness.

Prayer: I thank You, Lord, that even though sin rules the world, there is some relief from unrelenting selfishness. Forgive me for my coldness and help me to look to Your example so that I can be more selfless, even with those who are selfish. Amen.

Ref: "Ants can Learn to Favor Friends over Family." *Discover*, May 1986, pp. 10-13.

Synchronized Fireflies

Psalm 119:105
"Thy word is a lamp unto my feet and a light unto my path."

Fireflies in Southeast Asia regularly put on a show like that seldom seen from fireflies in the rest of the world. These fireflies have an ability found only among humans.

As the evening darkness deepens along a Southeast Asian riverbank, fireflies in the trees over the riverbank begin to flash. At first the flashing is familiar to anyone who has ever seen fireflies anywhere in the world. Flashes come randomly from different parts of the tree. But within a few minutes, something strange begins to happen. Instead of the flashes appearing randomly, they now appear in groups. Fireflies in one part of the tree flash in unison, then a group in another part of the tree flashes. In less than half an hour, the entire swarm, which may extend over more than one tree, is flashing once per second in perfect unison. This ability to pick up a rhythm to synchronize actions is found otherwise only in human beings.

Scientists report that different species have different rates at which they synchronize. They have also discovered that fireflies of one species ignore the synchronized flashing of other species. Some firefly swarms even flash in waves, creating an especially eerie sight along riverbanks.

One scientist who has been trying to understand how fireflies do this said that this behavior is so complex he has no idea how it works. Whether or not these scientists believe in the Creator, their wonderment is a tribute to His unsurpassed handiwork, even in the lowly firefly.

Prayer: Dear Father, there is no end to the wonders You have created in our world! How then can I ever understand all the spiritual wonders waiting in eternity with You. Even as I am cleansed by Christ's blood, teach me by Your Word. Amen.

Ref: Ivars Peterson. 1991. "Step in time." *Science News*, Vol. 140. August 31, pp. 136-137.

An Ancient Concrete Floor

Genesis 2:19
"And out of the ground the LORD God formed every beast of the field, and every fowl of the air; and brought them unto Adam to see what he would call them: and whatsoever Adam called every living creature, that was the name thereof."

A large, sophisticated concrete floor has recently been discovered in China. The floor dates to a period that, not long ago, evolutionists called the Stone Age.

The concrete floor, discovered in northwest China, extends the record of technological man far beyond any age imagined by evolutionists. Concrete of excellent quality was widely used in the Roman Empire. However, according to the evolutionary scenario, when the Chinese floor was built, man had just perfected stone axes and arrowheads.

The concrete floor measures 144 square feet. The concrete itself is a greenish black. Scientists report that when struck with an iron object it gives the same hollow sound as modern concrete. The concrete is similar to today's and includes silicon and aluminum. Sand, stone, broken pottery, and bones were mixed with the concrete. Obviously, so-called "Stone Age" man was involved in much more industry than some scientists had thought!

Scientists who believe the Bible's view of history are not surprised by this discovery. They know that every human being, from the very first one, has been completely human and quite intelligent. On his first day of existence Adam was smart enough to understand the animals and give them meaningful names! To a man with such understanding of God's creation, concrete would be a simple technology. The question really is, are we as smart today as Adam, who knew the Creator face-to-face.

Prayer: Dear Father in heaven, I thank You for the gift of intelligence. Though our day prides itself on knowledge, compared to Adam, my knowledge of You is poor. Help me to make Your Word the only true guide in my life. Amen.

Ref: L. Smart. 1986. "Scientists find ancient concrete." *The Herald* (New Britain, CT), January 25, p. 11.

Hot Sharks

1 John 4:19
"We love him, because he first loved us."

We've all been taught that fish are cold-blooded. In many people's thinking, perhaps the most cold-blooded of all is the shark. Now scientists have discovered that four types of sharks can raise their internal body temperatures. This fact offers a powerful witness to the love and provision of the Creator.

There is good reason for sharks to be hungry nearly all the time. They seldom catch a large tuna or other creature to meet their needs. For this reason, a shark can live for weeks on the fat reserves stored in its liver. Amazingly, the shark's liver can account for up to 20 percent of its body weight. So, when feeding is good, the faster a shark can digest, the more food it can store for lean times.

The shark increases its body temperature to speed digestion. When food is plentiful, a great white shark can increase its body temperature above that of the surrounding water by over 13 degrees Fahrenheit. Normally the heat generated by the shark's muscles is lost mainly through the gills. When the heat is needed to speed digestion, however, a complex radiator called a "miraculous web" moves into action. The miraculous web transfers the heat from the blood going to the gills to the blood coming from the gills. As a result, the shark can more efficiently digest and store food.

The shark seems unlovable to us. However, its Creator lovingly designed special features such as the "miraculous web" into the shark because He loves the creatures He made. After all, He loved us and sent His Son to die for our salvation when we were unlovable.

Prayer: Dear Father, You loved me and sent Your Son for my salvation even though I had nothing worthwhile to give to You. I thank You for Your forgiving love. Help me to love others with that same kind of love. In Jesus' Name. Amen.

Ref: "The Fire in the Belly of the Beast". *Discover*, Feb. 1988, p. 8.

"Prehistoric Man" and the Space Shuttle

Isaiah 45:12
"I have made the earth, and created man upon it: I, even my hands, have stretched out the heavens, and all their host have I commanded."

Would the early Indian settlers in North America have been able to understand modern twentieth century America? Beneath the peat deposits of Windover Pond in Titusville, Florida, scientists are finding the bodies and possessions of Indians who lived in Florida thousands of years ago.

The bones of almost 100 individuals have been found buried in the peat at the deepest part of the pond. Each was buried in a near-fetal position with feet toward the east and head toward the west. Each body lay on its left side, facing north. Children were buried with tools and jewelry. To the surprise of scientists, each burial shows clear evidence of ritual and knowledge of the compass points. Textiles found at the site convinced scientists that this ancient society was much more advanced than they had ever expected.

Scientists also marveled at the discovery of a young person who had a crippling spine disorder. The fact that the child lived 15 or more years illustrated that this was not a survival-of-the-fittest type of society.

We who know that the Creator made man fully human from the start are not surprised to see the same humanity in ancient man as we know today. These people would have watched the space shuttle fly off into space with a universal human curiosity and desire to know more about the new frontier of space. Man has always been man because we were made that way by our Creator.

Prayer: Dear Lord, I thank You that I was created by Your action and that I was saved from the consequences of my sin by Your suffering, death, and resurrection. Help me to better live as You would have me to live. Amen.

Ref: Michael Connelly. 1986. "Raiders of the Black Hole." *Sunshine* (Fort Lauderdale, FL), Dec. 21, pp. 7-11.

Do You Have "Extra" Parts?

1 Peter 2:15-16
"For so is the will of God, that with well doing ye may put to silence the ignorance of foolish men: As free, and not using your liberty for a cloak of maliciousness, but as the servants of God."

I remember reading in grade school that the human appendix is a useless organ. My textbook said that scientists thought that the appendix was once used to help digest the tree bark that our supposed ape-like ancestors ate. Does the human body have "extra" parts? Do we have organs that we no longer use because we have evolved away from needing them?

Atheists have argued that our "useless" organs prove that we weren't created. No Creator, they said, would make useless organs that can threaten our health. Creationists responded that since our knowledge of the human body is incomplete, we cannot say that an organ is useless simply because we are ignorant of its job.

In the 1890s, scientists said that the human body has about 180 organs that are useless leftovers of our past evolution. As a result, doctors used to be quick to remove a child's tonsils. Today we know that our tonsils have several jobs and are an important part of our immune system. Our appendix has been found to serve as a backup for other organs. If your spleen is damaged, your appendix will take over some of its functions. Wisdom teeth are important in chewing food, especially when the diet includes more coarse materials.

The argument over whether we have extra parts left over from evolution has been completely won by creationists. Today, science recognizes that we have no useless leftovers from evolution.

Prayer: Heavenly Father, You know all things and we know so little. Help me to realize my error when I imagine that I know things about which I am, in fact, ignorant. Through Your Word, enlighten me with truth. In Jesus' Name. Amen.

Ref: Jerry Bergman, Ph.D. *Vestigial Organs-A Brief Summary of the Latest Research*, pp. 111-115.

The Jerusalem Department of Public Works

2 Samuel 5:8a
"And David said on that day, Whosoever getteth up to the gutter, and smiteth the Jebusites, and the lame and the blind, that are hated of David's soul, he shall be chief and captain."

When human beings think that they have found mistakes in the Bible they are eventually judged by their own words. Beneath old Jerusalem lies a complex, ancient water supply system.

Archaeologists have long considered it to be poorly designed. Worse, they said that the biblical account of David's capture of Jerusalem was in error since it mentioned water tunnels that didn't exist in his time. Now these archaeological and biblical critics have been proven wrong on both counts.

A detailed study of the tunnels beneath old Jerusalem reveals a cleverly designed, sophisticated, dual water supply system that served the city. Nearly 3,000 years ago, ancient engineers modified a natural network of channels and tunnels under old Jerusalem to create the unique system. New tunnels were used to link existing natural tunnels so that an ambitious engineering project was completed with the least amount of work.

Research also shows that there are two openings to the system outside the walls of Jerusalem. It was one of these natural openings that David used to enter and conquer Jerusalem. Since we now know that they were natural and not built after David's time, no one can claim that the Bible's account is untrue.

Skeptics have claimed they have discovered many errors of historical or scientific fact in the Bible. Every time more facts have been learned, the Bible has been vindicated.

Prayer: Your Word is truth, O Lord! While I do not doubt Your Word in the Bible, I confess that I show that I doubt its importance by neglecting to study it as I ought. Forgive me for Jesus' sake and make my heart burn to study the Bible. Amen.

Ref: B. Bower. 1991. "Jerusalem Yields 'Natural' Waterworks." *Science News*, Dec. 7, p. 375.

Bigger Than Tyrannosaurus!

Psalm 143:5-6
"I remember the days of old; I meditate on all thy works; I muse on the work of thy hands. I stretch forth my hands unto thee: my soul thirsteth after thee, as a thirsty land. Selah."

All of us learned in school that the fierce *Tyrannosaurus rex* was the largest of the meat eaters. This dinosaur was large enough to reach into a second story window with his five-foot-long head and grab a person with his six-inch-long teeth. Longer than a railroad boxcar, he probably weighed about ten tons!

Nor was *Tyrannosaurus* a gentle giant. Healed and unhealed injuries on one fossilized *Tyrannosaurus's* bones showed that it had suffered several serious attacks at different times in its life.

Crocodiles are among the last living members of the dinosaur family. Recent excavations show that not so long ago, a meat-eating reptile even more massive than *Tyrannosaurus* terrorized a region of South America. The creature was an extinct species of crocodile called *Purussaurus*. The bones were found along the border between Peru and Brazil.

Imagine a crocodile that stands eight feet high and is 40 feet long! It could have easily picked up a black bear in its mouth and walked off with it. A second skull has been discovered that suggests that *Purussaurus* could have reached over 45 feet in length! It is thought to have eaten rodents that were the size of small cattle.

As we look into the past we see that there were once many more kinds of creatures, and on the average, they were larger and stronger than creatures today. That's exactly the pattern we would expect, based on the Bible's story of creation.

Prayer: Dear Father, I ask that You would use my voice, along with the fossil record's story of past death and present degeneration, to call more people to Your forgiving love toward us in Your Son, Jesus Christ. Amen.

Ref: Richard Monasters Key. 1991. "A Tyrannosaurus' Troubled Past." *Science News*, November 9, p. 303.

The Green Lacewing "Wolf"

Matthew 7:15
"'Beware of false prophets, which come to you in sheep's clothing, but inwardly they are ravening wolves.'"

Whether we look at the plant world, animals, or spiritual matters, predators often seek to disguise themselves as harmless to their victims. This is an effective strategy, often leaving victims unaware of danger until they are consumed by it.

Ants shepherd and protect some species of aphids that produce a tasty nectar. Shepherd ants are selected from the ant colony to herd the colony's tiny milk cows to good feeding areas. The slow-moving aphids are easy to herd. Any insect that approaches too close to the herd is warned away. Since aphids are considered a good meal by many insects, the shepherd ants are especially aggressive in protecting them against recognized predators.

The larvae of the green lacewing especially enjoy a good meal of aphids. The larvae actually disguise themselves to infiltrate aphid herds. They cover themselves with the same wax fibers that are produced by the aphids. Once the larvae look like aphids, they sneak into the herd unnoticed and are able to feed without gaining the attention of the shepherd ants.

While plant and animal predators can only kill their victims, spiritual predators can destroy our eternity. This is why the Lord warned us that false teachers often take the appearance of good spiritual people. He also told us how to identify them. While false teachers can take on a harmless or even good appearance, they can only *pretend* to be good. Ultimately, the way they conduct their lives and treat others will reveal what they truly are inside.

Prayer: I thank You, dear Lord, that You have told us how to identify false teachers who could destroy us. I ask that You would be my Good Shepherd and Protector from all harm. Amen.

Ref: Natalie Angier. 1984. "Thomas Eisner: The Bug Man of Ithaca." *Discover*, Feb., pp. 49-58.

Sunlight on Global Warming

Matthew 24:30
"And then shall appear the sign of the Son of Man in heaven: and then shall all the tribes of the earth mourn, and they shall see the Son of Man coming in the clouds of heaven with power and great glory."

The world is a fearful place for those who don't fear, love, or trust the Lord. Such a world is filled with demons of our own creation. Those demons might inhabit a jungle bush or the sky above a skyscraper. They are found in cultures that cook with campfires and those that cook with microwaves. Wherever these demons are found, they demand man's fear and must be appeased.

Who remembers over 20 years ago when scientists feared that the Earth was entering a new ice age? It was not long after this fear was proven to be unfounded that they began to see a new demon, global warming. While some insist that global warming is well underway, most admit that presently there is no evidence the Earth is warming. They warn though, that it could start at any minute and it will be caused by man's pollution.

A recently released study, however, shows a strong relationship between the sun's energy output and land temperatures. Both have been recorded since the late nineteenth century, but have never before been compared with each other. Of course, if the sun's natural variability is the cause of temperature variations, there is little we can do about it.

It is folly to think that man is so powerful he can change the world's climate. But the Bible's wisdom tells us that there is a Creator Who controls all things, and it tells us how He has loved us. It also tells us that this Earth will be inhabited when He returns to it.

Prayer: *Dear Father in heaven, I commend into Your loving hands all of my fears that have been generated within me by the terrified world around me. Let me, comforted by Your love, be a witness of hope in You to those around me. In Jesus' Name. Amen.*

Ref: "Global Warming: Checking out the Sun." *Science News*, Vol. 140, p. 380.

Robot Bugs

Psalm 8:1
"O LORD our Lord, how excellent is thy name in all the earth! who hast set thy glory above the heavens."

The nervous system of an insect is infinitely more complex than our most sophisticated computer-driven robots. Robots can only solve the simplest problems. While an insect has no difficulty walking, a walking robot is limited to an uncluttered, flat floor where it clumsily clanks about until something goes wrong.

An Indiana high school student recently had a better idea. He collected syringes for pistons and voltage converters from old computer printers, along with other spare parts. If he had simply assembled them and wired the result into his computer, he would have done well to produce the usual, clumsy robot walker. In an effort to build a better walking robot, he wired crayfish nerve cells that control walking between the robot and his computer. His computer acted as the brain while the nerve cells processed the signal as a real animal would. His robot walked more like a living creature and earned him national attention.

As a result of the student's success, researchers have built robot insects with simple, computer-simulated neural networks like those in the crayfish nerve cell. The result has been robot bugs that walk much more like real insects, although not well enough to fool anyone.

Such is the excellence of the work of the Creator. A few cells from a crayfish's nervous system can do so much more than all of our most sophisticated computers and programs!

Prayer: *Lord, Your work is excellent in every aspect. I thank You for that, and I pray that through Your Holy Spirit You would help me to be excellent at everything I do. Let all my actions be for Your glory and not for mine. Amen.*

Ref: Elizabeth Pennisi. 1991. "Robots Go Buggy." *Science News*, Vol. 140, Nov. 30, pp. 361-363.

Multi-Legged Chemical Warfare

Psalm 68:5
"A father of the fatherless, a judge of the widows, is God in his holy habitation."

Insects use a vast array of chemical strategies to defend themselves from predators. Some of their strategies are so ingenious that they could only have been made by an all-wise Creator Who knows how everything in the creation works. One of the world's greatest experts on insects has noted that insects are far more complex creatures than biologists had ever imagined.

Most chemical warfare used by insects is for defense. The goal of most of this defense seems to be to make the insect unpleasant to a predator. Millipedes, beetles and other insects spray an irritating chemical called quinone at their enemies. If an ant is shot with quinone, it becomes dizzy and staggers away while trying the wipe the chemical off itself. Some millipedes spray cyanide at intruders to discourage them. The whip scorpion fires a stream of liquid that is 84 percent acetic acid at its enemy. The female assassin bug collects camphor from certain plants and spreads it on her eggs to repel predators.

One beetle is so full of the birth control hormone progesterone that it is a walking birth control pill. If you ate a steady diet of this beetle, you would never produce any children. The chemical defenses of insects are both biologically sophisticated and far-sighted.

The Creator has given some of His creatures chemical weapons for their defense. To others He has given speed or intelligence. Throughout, it is clear that He has given His love to all His creatures.

Prayer: You have given many different kinds of gifts to Your creatures, Lord. I thank You that You have given Your love to all Your creatures. Help me to use the gifts You have given to me to let others know about Your forgiving love. Amen.

Ref: Natalie Angier. 1984. "Thomas Eisner: The Bug Man of Ithaca." *Discover*, Feb., pp. 49-58.

Do We Always See Clearly?

Matthew 7:3
"And why beholdest thou the mote that is in thy brother's eye, but considerest not the beam that is in thine own eye?"

All of us have a blind spot where the optic nerve enters the retina of the eye. Yet, we don't see a hole in our field of vision. Scientists always thought that this was because our brain simply ignores the ever-present blind spot. New research is showing that an even more complex system erases that blind spot.

Researchers used computer-generated images to create artificial blind spots in volunteers' field of vision. These blind spots – sometimes round, sometimes square – were shown against various moving backgrounds. Different shapes and backgrounds were mixed with different colors and sometimes moved slightly in the hope of learning whether the eye, the brain, or both, process that blind spot out of our vision.

Studies are continuing to test researchers' conclusions. Currently, they believe that cells in the brain fill in our blind spot by duplicating the information immediately surrounding it. Separate mechanisms fill in color and texture. These separate mechanisms may even be in different parts of the brain. Strangely enough, though apparently invisible, objects completely in the blind spot seem to be perceived in some unknown way.

Just as the brain and eye work together to make us blind to a true blind spot, sometimes the world, the devil, and our flesh work together to make us blind to our own faults. How reassuring to know that we can confidently believe that Jesus Christ carried our sins for our justification rather than permitting us to justify ourselves.

Prayer: *I confess to You, my crucified and risen Savior, all my sins, especially my ability to see sin in others more clearly than in myself. Forgive and enable me to see more clearly my own faults so that I may bring them to You. Make me more blind to the faults of others. Amen.*

Ref: B. Bower. "Vision System Puts Eyesight in Blind Spots." *Science News*, Vol. 139, p. 262.

The Strange Case of the Singing Fish

John 1:3
"All things were made by him; and without him was not any thing made that was made."

During the mid-1980s, the residents at the north end of San Francisco Bay began complaining about a strange droning noise coming from the Bay. During the months of July and August, the odd noise started after sunset and continued until sunrise. People living on houseboats found that the noise disrupted their sleep.

Puzzled and tired citizens began a search for the source of the noise. Sophisticated acoustical maps were made of the bottom of the Bay. The local sewage treatment plant was ruled out. A research lab operated by the Army Corps of Engineers was also ruled out. And the hum continued through the long nights.

Eventually, biologists joined the search. Soon, the toadfish, also known as the singing fish, was identified as the source of the sounds. The droning noise was actually male fish offering their mating call to attract females. All fish have gas or air bladders that act in the same way as a submarine's ballast tanks. The gas bladder adjusts to pressure changes, thereby helping fish stay at the depth they desire. A few fish, including the toadfish, have a set of muscles that rapidly vibrate the air bladder. The bladder then acts as a resonating chamber, and the sound is transmitted into the water.

The Bible tells us that all things were created by the same God Who was made flesh for our salvation. Since communication is part of God's nature, we should not be surprised to find communication so universal among the creatures He made.

Prayer: I thank You, Lord God, that communication is part of Your nature and that You have communicated to us. Forgive me for neglecting the study of Your Word to me in the Bible, and help me to be more eager to study it. In Jesus' Name. Amen.

Ref: John E. McC. 1986. "In Sum, it was Some Hum." *Discover*, June, pp. 67-71.

Horses Before Dinosaurs

Genesis 1:24
"And God said, Let the earth bring forth the living creature after its kind, cattle, and creeping thing, and beast of the earth after his kind: and it was so."

According to evolution, dinosaurs developed tens of millions of years before horses and were extinct long before the first horse galloped across the countryside. According to the Bible, horses and the land dinosaurs were made on the same day. This means that it should be possible to find evidence of horses in the same rocks as we find dinosaurs. Such a discovery would be a challenge to evolution!

The scientific literature has reported the existence of several examples of apparent horse hoof marks in rock that is supposedly older than most of the dinosaurs. Some of these apparent hoof prints have been found in the Pennsylvanian and Permian rocks of the Grand Canyon. Similar prints have also been reported in Triassic rock in Connecticut and in the New Red Sandstone of Scotland.

Evidence of horses living before most of the dinosaurs lived, or even at the same time, damages the theory of evolution beyond repair. Paleontologists gave the creatures that may have made these tracks a variety of names. The names frequently contained the word "equus," which means horse. But this important information has been withheld from the textbooks. Worse, modern evolutionists, aware that they are in serious trouble, have tried to deny the work of earlier paleontologists.

What upsets modern evolutionists most about these tracks is that the tracks uphold the Bible's account of the order of creation!

Prayer: Dear Father in heaven, while I don't need science to know that Your Word is truth, I thank You that even the rocks speak in support of Your truth. Grant me the words to speak Your truth to those around me. In Jesus' Name. Amen.

Ref: Edwin D. McKee. 1982. "The Dupai Groups of the Grand Canyon." *Geological Survey Professional Paper*, p. 93.

Racing Cockroaches

Romans 1:20
"For the invisible things of him from the creation of the world are clearly seen, being understood by the things that are made, even his eternal power and Godhead; so that they are without excuse:"

Scientists have been studying the different ways in which insects move, hoping to discover a better way to design mobile robots. In doing this, they have inadvertently recognized the wisdom and creativity of the Creator, even though they may not care to admit His existence.

Insect locomotion uses a variety of methods to travel over terrain even humans could not cover. Think of how the crab scuttles sideways across the beach. The spider walks up the wall and across the ceiling. Some insects walk on water. Then there's the inchworm. Stomatopods move by curling up and rolling away backwards. To their amazement, scientists have found that whether a creature has two, four, six, eight, or more legs, legs are always designed to attach to the body in a way that puts the least stress on them.

Many insects can change their method of locomotion to adapt to different situations. High-speed photography has shown that startled cockroaches race off by standing and running on their two hind legs only. In this position, they can reach speeds of over three miles per hour!

Because scientists find the wisdom of good design in nature, they search to see how nature has solved the engineering problems they face. If everything was truly the accidental consequence of chance, wouldn't scientists be studying accidents to find solutions to their problems? Though they may be embarrassed to admit it, by looking to nature, scientists are recognizing the wisdom of the Creator!

Prayer: I thank You, Lord, that I am fearfully and wonderfully made. Forgive me for the times that I have failed to see Your wisdom and power in my everyday life. Help me to be more trusting of Your promise that You are always with me. Amen.

Ref: E. Pennisi. 1991. "Scoot, Scramble and Roll." *Science News,* Vol. 140, Nov. 30, p. 363.

Is Your Brain Really Necessary?

Deuteronomy 6:5
"And thou shalt love the LORD thy God with all thine heart, and with all thy soul, and with all thy might."

How does the brain remember what it learns? What happens when you think? Is your mind the same as your brain?

As modern science learns more about the wonders of the brain, it also learns that it is farther than ever from learning how the brain works. Some former materialists are beginning to ask whether man has a non-material, spiritual, part.

A few years ago, doctors tried to save a child suffering from a severe brain disease by removing the entire left half of his brain. Since important centers for language and speech are in the left half of the brain, they did not expect the child to speak in a normal way ever again. Not only did the child recover, but as he grew into adulthood, his language skills were far above normal. Professor John Lorber writes about a student who had an IQ of 126 and held academic honors in mathematics, yet was found to have virtually no brain tissue at all. Professor Lorber decided to study similar cases. He found several, and at least half of these people who had 95 percent of their cranium filled with spinal fluid instead of brain tissue had higher than average IQs. Professor Lorber wrote about these cases in an article entitled "Is Your Brain Really Necessary?"

Yes, our brains are really necessary. These examples show us that the brain God has given us is even more marvelous than we thought. They also show that there is a great deal more to us than simply tissues and organs. Not one of us can claim that God didn't give us the mental abilities to be productive in glorifying Him!

Prayer: Father, I thank You for the mental gifts and abilities You have given me. Forgive me for Jesus' sake for not developing and using them as well as I could have in Your service. Help me to love You with all my heart and all my mind. Amen.

Ref: Paul D. Ackerman. 1990. *In God's Image After All: How Psychology Supports Biblical Creationism*, pp. 68-73.

A Spider Treatment for Stroke?

Mark 2:17
"When Jesus heard it, He saith unto them, They that are whole have no need of the physician, but they that are sick: I came not to call the righteous, but sinners to repentance."

A spider's venom paralyzes its victim, keeping the victim fresh, yet immobile. Spider venom does this by slowing or stopping the work of a chemical called glutamate, which controls muscle movement in insects. Glutamate is also an important messenger chemical in the human brain.

When someone has a stroke, the chemistry changes in those brain cells that are not receiving enough blood. Under these conditions, glutamate actually kills brain cells. So medical researchers have been searching for medically safe drugs that can be given to stroke patients to block the damage inflicted by glutamate.

Neurobiologist Hunter Jackson is searching for drugs in spider venom that might be safely used to block glutamate's damage. To do this, he collected, raised, and milked spiders. He had to do 10,000 milkings to get enough venom to study. Spider venom is a mixture of many different chemicals. Jackson says that some of those that block glutamate have not proven safe. But animal tests now show that about 20 of the chemicals from spider venom deserve more study for possible use.

We can easily be tempted to try to find the solutions for all our problems within this creation. We need to remember that while these solutions are blessings, the solutions found in the creation, at best, only solve temporary problems. In contrast, forgiveness of sins through Jesus Christ provides a permanent solution to the problems caused by sin.

Prayer: Forgive me, Lord, for those times that I have looked for temporary earthly solutions for problems that are really caused by my sin. Enable me to do a better job of looking to Your suffering and death for the forgiveness of my sins and to Your resurrection for my new life. Amen.

Ref: Elizabeth Pennisi. "Spider Toxins May Take Bite Out Of Strokes." *Science News*, Vol. 139, p. 270.

Do You Have a Bad Attitude?

Matthew 24:12-13
"And because iniquity shall abound, the love of many shall wax cold. But he that shall endure to the end, the same shall be saved."

Do you think that people are generally out to benefit themselves at your expense? Or are you someone who usually makes sure that you get what you feel is your share, even if it's at someone else's expense? Research shows that our mental attitudes can shorten our lives.

Researchers studied the attitudes of 500 adults in 1969. They were asked about their agreement with statements like, "In a time of crisis, people will generally look out for themselves." Then, based on their answers, they were rated on a scale to show how suspicious they were of others. Researchers kept track of the subjects for 15 years. By 1984, 143 of the 500 had died. Known causes like accidents were ruled out. The result showed that the more suspicious a person was, the more likely he or she was to be among those who had died. The differences in survival rates were surprisingly large.

Researchers suggest that the higher survival rate of the less suspicious is related to their ability to form more and closer relationships with others. Those who have close friends live longer and less stressful lives, probably because they have the support of others.

The Bible correctly teaches us that all human beings are by nature sinful and selfish. It also teaches us that the God-pleasing response to this fact is not suspicion, but forgiveness. God's will for us is always the healthiest and happiest way to conduct our lives.

Prayer: Lord, I confess that I have been suspicious when I should have been forgiving. While I ask that You protect me from ignorantly getting into a dangerous situation, I also ask that You would teach me not to replace forgiveness with suspicion. Amen.

Ref: "Only the Hostile Die Young." *Discover*, February 1988, p. 15.

Voodoo Aspirin

Genesis 1:11
"And God said, Let the earth bring forth grass, the herb yielding seed, and the fruit tree yielding fruit after his kind, whose seed is in itself, upon the earth: and it was so."

When they flower, some plants actually generate their own heat. The voodoo lily can raise its temperature by 25 degrees. As it does so, it releases a scent that attracts the beetles that pollinate it.

When the voodoo lily generates heat, it burns starch in a way similar to our own bodies. Scientists weren't sure how voodoo lilies generated their heat and scent until only a few years ago. Once they unraveled the complex chemical reactions going on within the plant, they found unexpected surprises. They discovered that a component of aspirin is responsible for raising the lily's temperature. They remain puzzled about why salicylic acid should *lower* our temperature, yet *raise* the temperature of the plant.

Scientists were also amazed when they discovered how many separate complex chemical reactions take place to produce a chemical symphony that helps the lily reproduce. Not only does the lily make its own aspirin, but also it makes the same chemicals that are found in the rotting meat that its pollinator beetles feed on. The rising temperature of the plant evaporates the chemicals, creating a scent that attracts the beetles. As the beetles look for food in the flower, they become covered with pollen, which will be delivered to the next lily they visit.

The voodoo lily is an example of how, even in the plant kingdom, there is no such thing as a simple life form. As this plant duplicates several of the processes taking place in animals, we see that all life was created at about the same time by one Designer!

Prayer: Dear Lord, everything You have made is a cause for wonderment and thanksgiving. I confess that I have not given You the thanks that I should have. Forgive and help me to be more thankful for all Your goodness to me. Amen.

Ref: "Do Do That Voodoo That You Do So Well." *Discover*, Feb. 1988, p. 10.

Butterfly Physics and Stealth

Psalm 104:27
"These wait all upon thee; that thou mayest give them their meat in due season."

One would not think that a brightly colored, relatively slow butterfly has much of a chance against a bird moving in for lunch. But you would be surprised. Scientists were.

Some butterflies are protected by distinctive, bright markings that are poisonous. What about those butterflies that are bright *and* tasty?

Tropical birds that eat butterflies are usually fast and very agile fliers. They often have long, pointed beaks well-designed to catch butterflies. It would seem that butterflies wouldn't have a chance against faster-flying birds. In truth, it's the bird that has the greatest disadvantage. The muscles that control the tasty butterfly's wing are relatively stronger in terms of lifting power than a bird's muscle. In addition, tasty butterflies have about 50 percent more flight muscle than the poisonous butterflies that birds avoid. Since butterflies weigh so much less than birds, they can change direction more rapidly. So even though a bird can fly much faster than a butterfly, by flying in a tight, darting, erratic pattern, a skillful butterfly can easily escape its predator.

Not only is the butterfly wonderfully designed, but also it's designed in view of the birds that might eat it. Notice how God carefully countered the bird's advantage of speed by giving the butterfly stronger muscles and a more maneuverable body. Darwinism can only pretend to explain that!

Prayer: I thank You, Lord, for the beauty of both the butterfly and the bird. I also thank You that You are a personal God Who cares about everything You have made. Fill my heart with the same love toward all You have made. Amen.

Ref: James H. Marden. "Newton's Second Law Of Butterflies." *Natural History* 1/92, pp. 54-60.

How to Freeze a Turtle

Job 5:8-9
"I would seek unto God, and unto God would I commit my cause: Which doeth great things and unsearchable; marvelous things without number:"

The painted turtle is found further north than any other turtle in North America. During its first year of life, a painted turtle survives temperatures as cold as 18 degrees Fahrenheit.

In mid-June, painted turtles begin to lay their eggs. Each nest holds from seven to nine eggs. Some females will make two nests. The eggs are buried, safely out of sight of predators, and the mother turtle returns to her normal habitat. The young hatch in ten or eleven weeks. After hatching, they remain buried in the ground, and therefore safe from predators, all winter. The problem is that turtles freeze solid at the temperatures found at nest depth in the winter. Usually, when living cells freeze, the long, sharp ice crystals that form in them puncture the cell membrane, killing the cell.

As the baby turtles freeze, even the heart and brain eventually freeze. There is no breathing and no heartbeat. Only a tiny bit of electrical activity in the frozen brain reveals that life remains in the body. Why don't ice crystals rupture the cells? The young turtle's liver makes special proteins that are circulated to every cell in the body. These proteins ensure the formation of very small ice crystals that cannot puncture delicate cell walls.

Only God could have invented such a unique method of protecting tiny, painted turtles. Even scientists marvel at this.

Prayer: Father in heaven, truly Your wisdom and power know no limit. I ask that You would help me to realize how Your wisdom and power are most clearly revealed in Scripture so that I may return to the Bible on a regular basis to marvel at Your works and glorify You. In Jesus' Name. Amen.

Ref: Janet M. Storey and Kenneth B. Storey. "Out Cold." *Natural History.* January, 1992, pp. 23-25.

An Inside Job

Luke 6:31
"And as ye would that men should do to you, do you also to them likewise."

The white-fronted bee-eater is an East African bird that lives in clans of up to 14 members.

White-fronted bee-eaters have several problems to deal with. Since they nest on the cliffs overlooking river banks, youngsters need a great deal of attention until they learn to fly. Once they are on their own, the young birds are often put to work by their parents as helpers. A father may even drive away his sons' mates in order to keep them as helpers.

Helpers bring food for their mother and brothers and sisters. They also clean the nest area and watch for danger. The most important job, however, is guarding the nest at egg-laying time. This is necessary because a female who doesn't have her own nest will sneak into another bee-eater's nest and lay her eggs there. If the eggs are laid before the owner of the nest starts laying her eggs, she will simply toss out the foreign egg. If there are eggs already in the nest, she will also care for the foreign eggs. The important job of guarding the nest is usually given to a daughter. But scientists have observed that sometimes it is the daughter, while on guard duty with the mother absent, who sneaks into the nest and adds a few eggs of her own!

In His goodness, the Creator has given bee-eaters a way of life in which helping each other is part of their nature. This kindness toward each other improves the quality of life. It can serve as an example to us that the world is not designed to favor the survival of the most selfish or aggressive.

Prayer: Lord, in a world that sees only personal loss when kindness is offered, help me to remember that You gave me all things freely. Help me to see the reward of good will in helping others, even those who cannot help me in return. Amen.

Ref: Bruce Fellman. 1992. "Looking Out for Number One." *National Wildlife*, Dec.-Jan., pp. 46-49.

Babylon Rent-A-Wagon

Acts 7:4
"Then came he out of the land of the Chaldeans, and dwelt in Charan: and from thence, when his father was dead, he removed him into this land, wherein ye now dwell."

Today, people regularly travel a thousand miles in modern airliners. A thousand miles is a long day's drive, but it can be done. We would never think of walking or riding a horse that far.

It was out of this questionable sense of superiority that critics of the Bible rejected the accuracy of the Bible's record of Abram's travels. According to the Bible, when he lived in Ur, in Chaldea, God instructed him to travel where God would lead him. That trip took Abram from northern Mesopotamia to Haran. After the death of Abram's father, God finally led him on to Canaan. The total length of Abram's travels amounted to more than a thousand miles. Abram could not have travelled that far so long ago, critics argued.

Several years ago, a clay tablet was discovered in ancient Babylon that sheds some light on the critics' challenge. The tablet was a wagon rental contract. The contract forbids the wagon renter from taking the wagon as far as the lands of the Mediterranean coast. One archaeologist observed that this meant that travel over such great distances was so common that Babylon's rent-a-wagon outlet had to place mileage limitations in rental contracts.

Over the years, hundreds of challenges have been raised against the accuracy of the Bible. Yet, none of the Bible critics' words have stood the test of truth. The Bible – God's Word – has!

Prayer: Dear Father, the Word made flesh for my salvation, Your Son Jesus Christ, is revealed to me on every page of the Word written for my salvation. Help me seek truth only in Your Word and not in the words of men. In Jesus' Name. Amen.

Ref: Wayne Jackson. 1991. "Archaeology and Abraham's Journey." *Reasoning From Revelation*, Vol. III, No. 11. Nov., p. 22.

Reverse Engineering

Acts 17:11
"These were more noble than those in Thessalonica, in that they received the word with all readiness of mind, and searched the scriptures daily, whether those things were so."

The human brain is like a network in which each brain cell is connected to thousands of other brain cells. Scientists have long suspected that this unique design helps give the human brain its intellectual power.

The brain is so complex, however, that our most sophisticated mathematics are insufficient to cope with it. As a result, scientists are left with only trial and error as they try to study it in an effort to improve computers.

A new step forward was made recently when scientists designed an effective computer neural network chip. The silicon neural network crudely mimics the brain's neural network. The manmade networks are designed to learn all by themselves. Scientists wired a number of networks together and gave them a memory of the characteristics of three-dimensional images. They then exposed the neural network to images it had never seen before to see if it could compare the new images with its memory and tell whether the new images were three-dimensional. Scientists report that while it took the network a long time, it eventually learned to judge depth, even if it saw only part of an image.

Scientists admit that they are working from the brain's design to learn how to make better computers. As one of the scientists working on the project said, they are doing "reverse engineering." They are starting with the genius of our Creator, Who made the human brain, to learn how thinking and learning take place.

Prayer: Dear Father in heaven, all of the creation, including unbelieving scientists, glorify You, even if they don't know it. Help my praise to You be conscious, deliberate, and often. In Jesus' Name. Amen.

Ref: E. Pennisi. 1992. "Neural-net Neighbors Learn From Each Other." *Science News*, Jan. 11, p. 23.; Paula Rooney. 1990. "Neural Nets Breed Applications-But Not EE Employment." *EDN*, Oct. 18, pp. 105, 108, 120.

Ceramic Miracles

Psalm 9:1
"I will praise thee, 0 LORD, with my whole heart; I will shew forth all thy marvelous works."

Scientists who are working on newer and better materials out of which to make things are concentrating their attention on ceramics. Ceramics do not break or wear as easily as other materials, including metals. High temperatures don't cause them to weaken. Some ceramics have even shown the promise of bringing about another revolution in electronics. Ceramics have one major problem. They are difficult to form into usable parts. Sometimes they are also too brittle to work well as machine parts.

Nature, on the other hand, regularly custom-forms ceramics for the most demanding duties. The ceramics made in living bodies include teeth and shells. Living creatures bind inorganic crystals into a hard, custom-fashioned ceramic that is both strong and much less brittle than manmade ceramics.

Taking their cue from living things, scientists have successfully made stronger, less brittle ceramics patterned after biological ceramics. The new ceramics offer great promise in designing radical new electronics. Perhaps most dramatically, one version has served well as artificial bone. The body even accepts the ceramic as though it were true bone!

Modern science has gained some wonderful knowledge. Now, modern scientists have learned that to solve an engineering problem it is better to learn how our wise Creator solved the same problem in the first place.

Prayer: I thank You, Father, that You have given me my body with all its wondrously designed parts. Help me to take good care of it as a gift from You and raise it back to life in the resurrection of the dead. In Jesus' Name. Amen.

Ref: E. Pennisi. "Nature Points The Way to Tougher Ceramics." *Science News*, Vol. 140, p. 150.

The Panda's Thumb Revisited

Psalm 2:1
"Why do the heathen rage, and the people imagine a vain thing?"

Many who reject the Creator work hard to try to prove that He doesn't exist. Creationists argue that the creation is so well designed that it must have a Designer. After all, a watch is much simpler than a bumblebee. If a watch needs a designer and builder, so does a bumblebee.

Evolutionists argue that some things are not designed as well as they could be. This, they say, proves that there is no all-wise Creator Who designed and made everything. One example of what evolutionists claim to be a lack of design in the creation is the so-called "panda's thumb." Below the five toes on its paw, the panda has a sixth appendage. This so-called "thumb" is formed from an extension of a wrist bone and a pad. Evolutionists say that if the panda were truly designed by God, it would have a real thumb or none at all. But other evolutionists have pointed out that the panda's thumb design gives it the ability to handle bamboo with great precision – almost as well as a surgeon handles a scalpel!

The evolutionists' attempts to prove that there is no Creator cannot succeed. Negative statements cannot be proven. Simply because we don't know how a thing works or why it's built the way it is doesn't mean no one made it.

Man's most ambitious attempts to deny God cannot begin to deny Him, as the people of the former Soviet Union have surely learned!

Prayer: Father, I thank You for the increase in those who are turning to You and for those who call upon Your Name. Show me how I can do more to be a part of the spread of Your Word of truth in our needy world. Amen.

Ref: Bert Thompson, Ph.D. 1991. "Evolution's "New" Argument - Suboptimality." *Reason and Revelation*, Vol. XI, No. 11, Nov., pp. 41-44.

Explaining Too Much

Philippians 2:3
"Let nothing be done through strife or vainglory; but in lowliness of mind let each esteem other better than themselves."

We are all familiar with evolution's battle cry of survival of the fittest. Another way of putting it is that evolution favors the selfish.

Nature is full of surprises. For generations, evolutionists have explained why most animals breed promiscuously rather than taking mates for life. They argue the more they breed, the more of their genes continue in the next generation. Preserving genes is said to be an important goal of evolution. In other words, evolution favors the selfish.

The problem with this explanation is that naturalists are discovering increasing examples of individuals who either wait longer than necessary to mate or never mate at all. That's not very selfish. Worse for evolution, these animals actually help other individuals rear and feed their young. For example, two-thirds of all acorn woodpeckers never leave home to raise their own families. That's not very selfish! How do evolutionists explain this?

Evolutionists said that the helpers pass up opportunities to breed because it gives them the evolutionary advantage of preserving their genes. Did you catch it? Acorn woodpeckers who go out and breed as fast as possible are favored by evolution because they are preserving their genes. Those who help at home and never breed are favored by evolution because it preserves their genes! When evolution uses the same explanation for opposite activities, it shows that evolution is not science. It's only making up stories.

Prayer: You know so well, Lord, how we sinful human beings will make up and even believe anything that will justify our beliefs or make us seem innocent. Help me to abandon excuses and rely solely on Your suffering, death, and resurrection for my vindication from sin. Amen.

Ref: Bruce Fellman. 1992. "Looking Out For Number One." *Natural Wildlife*, Dec.-Jan., pp. 46-49.

Western Acupuncture

Psalm 12:6
"The words of the LORD are pure words: as silver tried in a furnace of earth, purified seven times."

Because it is sometimes connected to New Age religion, acupuncture is often seen as something negative. Stripped of its pagan mumbo-jumbo, however, acupuncture has a sound medical basis.

Medical researchers have shown that acupuncture stimulates certain nerves to produce natural pain-killing chemicals. These natural drugs can produce the effect of a sedative or anesthetic and can even lower heart rate. Researchers continue to debate just how acupuncture does this. But there is little debate that the benefits are real.

When we think of acupuncture, we usually think of the East and its religions. But it appears that the East did not have a monopoly on acupuncture. Veterinarians today still rely on an old device that produces the same effects. The twitch has been used in the Western hemisphere since the Middle Ages. It's a hoop of metal or even a loop of rope that is tightened around a horse's upper lip. It quiets the horse and makes him appear to be sleepy. Research shows that the horse's heart rate will drop below normal, and levels of natural pain killers in the horse's blood will double. Even when given an injection, the horse's heart rate will barely reach normal. Was such knowledge once part of man's God-given knowledge about the creation?

Knowledge that is handed down by word of mouth can be lost or its message confused. We thank God that He did not depend on word of mouth, but on direct inspiration, to record His Word for us!

Prayer: I thank You, Lord, that it is not upon men that I depend for Your Word. Increase within me the desire to learn more of Your inspired revelation in Scripture and apply it to my life so that I might live to Your glory. Amen.

Ref: "It's Nature's Way of Saying, Whoa!" *Science 84*, Dec., p. 6.

The Animal That Lives Without Air

Psalm 105:5
"Remember his marvelous works that he hath done; his wonders, and the judgments of his mouth;"

When little or no oxygen is dissolved in the blood of most animals or humans, death quickly follows. The chemistry by which our cells generate energy to stay alive requires oxygen. When oxygen runs short, a second, temporary energy-producing system kicks in. This system doesn't need oxygen, but it produces lactic acid which can kill living cells. Therefore, this second energy system is an emergency measure, used only when seriously starved for oxygen.

In order to avoid this dangerous situation, many amphibians and reptiles that hibernate in ponds, lakes, and rivers have the ability to remove oxygen directly from the water through special tissue on their throats. The painted turtle and the mud slider, however, hibernate as close to the freezing line below the water as possible. As a result, they often hibernate in water where the oxygen is quickly used and have lived completely without oxygen for up to 120 days. How do they do it?

During the summer, the turtle stores large amounts of a carbohydrate that can provide energy without using oxygen. The turtle has a unique built-in system that detoxifies the poisonous lactic acid that is built up. As a result, the turtle can live for months with no oxygen in its blood or in the water around it.

This unique ability is based on sophisticated biochemistry. It's clearly not the result of a genetic accident. It can only be the work of the unique creativity of our Creator!

Prayer: Lord, there is nothing that is impossible for You. I thank You for Your love for me in carrying my sin to the cross and rising from the dead so that I might have new life. Let me never forget that Your great power is used in love for me. Amen.

Ref: Janet M. Storey and Kenneth B. Storey. "Out Cold." *Natural History 1/92*, pp. 23-25.

Proof of Humans and Dinosaurs Together

Genesis 1:31
"And God saw every thing that he had made, and, behold, it was very good. And the evening and the morning were the sixth day."

According to the story of life offered by evolution, dinosaurs became extinct 65 million years before man ever set foot on Earth. According to the Bible, all living things were made during a six-day period of time. This means that humans and dinosaurs walked the Earth at the same time. If evidence that they lived together could be found, evolutionary history would be very seriously challenged.

Scientists who believe in creation know that if human remains were found fossilized with dinosaur remains, evolutionists could not ignore the evidence. That evidence has been discovered in a giant fossil graveyard which has even been described in the scientific literature. Textbooks on evolution neglect to mention it for obvious reasons.

The layer is a 15- to 18-inch thick layer of phosphate rock in the southeastern United States. The rock is 65 percent phosphate, which means that the layer is made primarily of the bones of dead animals. The bones were clearly deposited by a huge flood - the deposit is at least as large as the Everglades! And mixed into the layer are bones of all sorts of animals that evolution says could not have lived together, including human bones and Hadrosaurus bones.

Did men of the generations recorded in the Bible see great herds of duckbill dinosaurs ranging the grasslands of the Earth? According to the Bible, they could have. Now geology has finally caught up with the Bible!

Prayer: Dear heavenly Father, there is no end to learning about the wonderful things You have created. Overcome the spirit of doubt that is natural to me and all people so that I might better live and witness my faith. In Jesus' Name. Amen.

Ref: John Allen Watson. 1991. "Phosphate Rocks/Bone Phosphates of South Carolina." *The Ark Today* VI:5, Nov./Dec., pp. 14-19.

Protective, Teaching Fathers

Ephesians 6:4
"And, ye fathers, provoke not your children to wrath: but bring them up in the nurture and admonition of the Lord."

The first scientific studies that showed essential differences between males and females, men and women, were not well received by some. And many Christians have resisted the teachings that men and women are by nature the same. Christians have traditionally understood the Bible to teach that males and females each have essential, honorable, but different roles to play.

Unfortunately, many Christians have been unsure how to describe those roles. As we study the scriptural portrayal of the family, especially in Ephesians chapter 5, a picture begins to emerge. All the scriptural examples of motherhood and fatherhood can be distilled into one basic idea. What we call a healthy maternal instinct and a healthy desire on the part of the male to protect and prevent danger are only two sides of the same coin.

An example of what this means was recently discovered in bluebirds. Scientists have found that bluebird fathers offer their nestling daughters twice as much food as their sons. Of course, the sons are never allowed to go hungry. Scientists were puzzled by this until someone suggested that by doing this, father bluebird was teaching his daughters, by example, how to select a mate. Female bluebirds are very fussy about selecting a mate. One of the most important things they look for in a mate is the ability to provide plenty of food for the next generation.

In an age of declining parenting skills, it's good to be reminded that our Creator has given human parents instruction in His Word.

Prayer: Dear Father, I thank You that You are a perfect Father. Fill Christian parents among us with love and patience. Most of all, fill them with the instruction of Your Word so that they may teach their children about You in both word and deed. In Jesus' Name. Amen.

Ref: K. A. Facklemann. 1992. "Bluebird Fathers Favor Pink over Blue." *Science News*, Jan. 4, p. 7.

An Ancient Cure for Malaria

Ecclesiastes 1:9
"The thing that hath been, it is that which shall be; and that which is done is that which shall be done: and there is no new thing under the sun."

The modern, educated person often tends to look down on previous generations. Many people seem convinced that if we look back 1,000 or 2,000 years, some measurable evolutionary improvement in man must have taken place over so much time.

It is this spirit that makes so many today feel that they can pass judgment on the Bible. "After all," they say, "it comes from an ancient, uneducated time." They add, "Why, back then people didn't even know the difference between historical fact and a legend designed only to teach a lesson." This superior attitude of history suggests that since the ancients didn't have medical labs as good as ours, they couldn't know something we don't.

This is why medical researchers dismissed a cure for fever that dated to 340 A.D. An ancient Chinese book of herbal remedies recommended that to cure a fever, one should soak a particular weed in about a liter of water and drink the resulting tea. In typical modern fashion, researchers had doubts that this cure would work. However, the weed named in the folk cure provides a compound that may be an even more effective treatment for malaria than quinines. It treats cases that quinines don't affect and produces no side effects.

While our modern age may be advanced in some things, it's definitely behind past ages in other things – like humility.

Prayer: *I confess, dear Father, that it is too easy for me to become caught up in the sense of superiority that is so common among people today. Forgive me for Jesus' sake. Let my boasting be in Christ and not myself or this modern age. Amen.*

Ref: "A New Old Remedy." *Discover*, August 1985.

The Gardener Bowerbird

Psalm 90:17
"And let the beauty of the LORD our God be upon us: and establish thou the work of our hands upon us; yea, the work of our hands establish thou it."

Bowerbirds are found only in New Guinea and Australia. There are many types of bowerbirds, each with unique habits. However, bowerbirds are best known for their love of bright colors and the great care they use in decorating their constructions. All this activity is carried out to attract a mate. Scientists say that the brighter a bowerbird's feathers, the less elaborate its constructions.

The gardener bowerbird wears some of the brightest feathers among the bowerbirds. This spectacular bird, about the size of a blue jay, was last seen in 1895. Then to his amazement, an ornithologist surveying the modern bowerbird population saw one the first day he started looking. As it turns out, everyone else had been looking in the wrong place for this mysterious bird.

The gardener bowerbird has a yellow front and brilliant orange crest that goes over the top of its head, all the way to its nostrils. It builds a fairly simple bower that resembles a four-foot-tall maypole surrounded by a three-foot-diameter mossy area. This area has neatly arranged piles of bright-colored fruit around it. To attract a mate, the male makes different sounds, some of which resemble a frog's croaking. He then holds brightly colored fruit in front of his chest to further impress the female.

The standards of beauty so universal among God's creatures testify that all of the creation is His handiwork.

Prayer: Dear Lord, I thank You that even though groaning under the burden of human sin, the creation still shows so much startling beauty. I ask that my life would always be illuminated with the perfect beauty of the forgiveness of sins through Your death. Amen.

Ref: C. Simon. "Legendary Bowerbird Thrives in New Guinea." *Science News*, Vol.120, p. 326.

Who Was Neandertal?

Romans 14:13
"Let us not therefore judge one another any more: but judge this rather, that no man put a stumbling block or an occasion to fall in his brother's way."

When we hear about the "cave man" and the "Stone Age", most people think of Neandertal man. (Popularly called "Neanderthal," the preferred usage today is Neandertal.) When we think of Neandertal, most people think of a large, brutish, hairy, bent-over creature that looks like a cross between an ape and a human. Who, in fact, was Neandertal?

The first Neandertal bones were found in a cave in the Neander Valley, near Dusseldorf, Germany, in 1856. Evolutionary anthropologists of the day thought that they had finally discovered the so-called "missing link" between apes and humans. So when they reconstructed the bones into what they thought Neandertal looked like, he was portrayed as a hairy brute that could not stand fully erect.

Since that first discovery, over 300 Neandertal skeletons have been discovered. Evolutionists today recognize that the first Neandertals that were found suffered from rickets and arthritis, giving their bones only a superficial appearance of being less than human. Today's science has reclassified Neandertal as Homo sapiens. It is said that given a modern haircut and clothes, he could walk down any street and not attract attention.

There is one more fact that speaks eloquently for Neandertal's humanity. Neandertal practiced religion. He buried his dead with flowers and sometimes other possessions. He cared for his sick and the weak. Clearly, Neandertal offers no support for evolution.

Prayer: I thank You, Lord, that the truth has finally been learned about Neandertal. I ask that the other so-called evidences that doubters use to deny You would quickly be shown to be weak crutches so that no one may find anything in science a stumbling block to faith. Amen.

Winged Warriors

Romans 10:15
"And how shall they preach, except they be sent? as it is written, How beautiful are the feet of them that preach the gospel of peace, and bring glad tidings of good things!"

We don't often associate aggressive behavior and the need to fight with butterflies. Rather, we think of bright, beautiful creatures who decorate nature and always seem to cheer people up.

Scientists have always assumed that the bright-colored wings of butterflies served the purpose of attracting a mate. However, new research on tropical butterflies calls that idea into question. Researchers carefully changed the colors of male butterflies' wings, even painting patterns on them that were wrong for the species. When it came time for mating, females treated the disguised males the same as other males. It appeared as though the females paid no attention at all to the colors and patterns on the males' wings.

What practical purposes do the patterns serve then? Scientists know that they help predators avoid poisonous butterflies. They know that certain patterns deter birds. But butterflies have a nasty little secret. Many male butterflies tend to be aggressive with each other and very territorial. Butterflies are so fragile that almost any injury in a fight would be fatal. Rather than fight, male butterflies assert themselves by showing their wings to each other in something like a ritualized combat.

A universal standard of beauty is not something evolution could have produced. Butterflies are lovely, living ornaments with which the Creator and Author of all beauty has adorned His creation.

Prayer: I thank You, dear Father, for the beauty that You have built into the creation that continues to brighten our lives, despite the ugliness of sin. I ask that I would always find the brightest and best beauty in the forgiveness of my sins through Your Son, Jesus Christ. Amen.

Ref: L. Langley. "Butterfly Colors: Alluring or Alarming?" *Science News*, Vol. 121, p. 38.

Chimps Learn Math

Psalm 16:2-3
"O my soul, thou hast said unto the LORD, Thou art my Lord: my goodness extendeth not to thee; But to the saints that are in the earth, and to the excellent, in whom is all my delight."

Many people think that man's intelligence is what sets him apart from the animals. This idea is built on the evolutionary theory that as the most recently evolved creature, man is the most intelligent. It is also built on the evolutionary idea that man has no such thing as a soul.

Researchers now know that just because chimpanzees are not able to talk, it does not mean that they have learned nothing from language training. Researchers at the University of Pennsylvania tested the mathematical abilities of chimps that had language training against chimps that had learned only to do simple jobs. All of the chimps without language training flunked their tests.

The chimp that had graduated from 18 months of language training, though, seemed to understand small numbers, fractions, and showed analytical skills. When asked to match equal numbers of different objects, the language-trained chimp was correct 91 percent of the time. Then the chimp was given wooden disk fractions. Half a disk represented one-half, a quarter of a disk, one-quarter, and so on. The chimp was then asked to match those fractions with glasses of water that were half full, a quarter full, and so on. The chimp scored 88 percent correct in this exercise.

Human math ability is far beyond the level of a chimp. Moreover, human beings not only have a soul but a soul that is made in the image of God and that sets us apart from the animals. The fact is that God gave us the ability to have a relationship with Him through His Son, Jesus Christ.

Prayer: I thank You, Father, that when I was lost to You and held captive by sin, death, and the devil, You sent Your Son, Jesus Christ, to rescue me. Help me to grow ever closer to You as I read, study, and apply Your Word to my life. Amen.

Ref: "Teaching an Old Chimp New Math." *Science News*, Dec. 5, 1981, p. 363.

Want to Be More Negative?

Psalm 5:11
"But let all those who put their trust in thee rejoice: let them ever shout for joy, because thou defendest them: let them also that love thy name be joyful in thee."

If you haven't been feeling cheerful lately, maybe you should have more "negativity" in your life.

It is said that there is an increase in accidents, murders, suicide, and fights when the Santa Ana winds sweep into southern California. Canadians say that the Chinook winds produce the same effects. Wherever these winds blow, those who experience them tell the same tales. Scientists suspect that these winds generate positive ions. Positive ions in the air make people feel irritable and less well. On the other hand, negative ions are said to make people feel more relaxed and happy.

The study of the effects of positive and negative ions on people got off to a poor start at the beginning of the 1800s. However, more recent scientific studies suggest that about 30 percent of the population is sensitive to negative or positive ions. When there are a lot of negative ions, people report feeling better, and they have better reaction times. Positive ions increase complaints about irritability and headaches. Reaction times are also increased. If you have ever felt extra alert and perhaps even a bit euphoric as a thunderstorm approached, you were probably reacting to negative ions. Waterfalls and beaches are also natural generators of negative ions. One of the best negative ion generators you can have in your home is an aquarium.

It's good to know, though, that however our environment makes us feel, we can always have joy because of God's love for us in our Savior, Jesus Christ.

Prayer: Heavenly Father, I confess that sometimes I allow the world around me to take from me the joy that should always be mine because of Your love for me in Christ. I commend all of my concerns into Your hands and praise You in joy. Amen.

Ref: Linda Garmon. 1981. "Something in the Air." *Science News*, Vol. 120. Dec. 5, pp. 364-365.

Oldest Known Religious Shrine Discovered

Acts 17:27
"That they should seek the Lord, if haply they might feel after him, and find him, though he be not far from every one of us:"

The fact that man is by nature a religious creature was underscored with the discovery of the oldest known religious sanctuary in the world. The shrine was discovered in northern Spain. Evolutionists say that the shrine was built by early Stone Age man. In terms of biblical history, the sanctuary was probably built by some of the first post-flood settlers in Spain.

Scientists declared the ancient structure a religious sanctuary based on three criteria. First, it is a large structure that required the effort and cooperation of many people to build. Second, it has features that are unnecessary for daily living. Third, the structure is associated with a supernatural being. Scientists noted that the stone floor of the sanctuary shows a great deal of wear, indicating that it saw a lot of use. The worship center included an altar made of a limestone slab weighing nearly a ton.

The shrine also had a stone sculpture of a head. The right half of the head is human and the left half of the head is a carnivore of some sort. Worshippers at the site had separate storage places for sewing needles and hunting tools. Spear points, animal bones, and shells were found in a trench in the sanctuary.

Man is undeniably a religious creature. We have been made by our Creator in such a way that we are dissatisfied until we have a relationship with Him. You, too, can have a relationship with Him through the forgiveness of your sins which was earned for You by His Son, Jesus Christ.

Prayer: Dear Lord, only in You can I be satisfied. I thank You that I can indeed be satisfied without fear because You have carried my sins to the cross and brought me new life through Your resurrection from the dead. Amen.

Ref: C. Simon. 1981. "Stone-age Sanctuary, Oldest Known Shrine, Discovered in Spain." *Science News,* Dec. 5, p. 357.

Why Are Human Fossils Scarce?

Psalm 18:30
"As for God, his way is perfect: the word of the LORD is tried: he is a buckler to all those that trust in him."

One question we are often asked is why more human fossils are not found in older rocks. In order to become a fossil, a creature must be buried rapidly in sediment so that it does not decay. Aquatic animals would be much more likely to produce fossils in, say, a worldwide flood than land animals; and indeed, geologists universally use small marine creatures as their "index fossils."

Usually, the smaller the creature is, the larger the population, and this is another reason that small marine fossils are so plentiful. They are found worldwide from almost the very lowest sedimentary rocks to those at the surface and even at mountaintops. Generally speaking, the birds and the larger mammals, including man, have much smaller populations and are indeed found much less frequently.

If all these deaths resulted from the Great Flood, then as the valleys filled with water, the larger animals would run for the safety of higher ground; small animals would be exhausted and drowned as would the birds who could not fly in the heavy 40-days of rain. Mankind would have survived until exhausted on floating vegetation. For the most part, the drowned bodies of birds and small mammals, including man, would decompose quickly in the swirling waters and not be trapped in sediment. The larger animals, such as the dinosaurs, would easily be trapped in mud and rising sediment and thus be good candidates for fossilization.

The pattern of fossils actually found worldwide does not support the theory of evolution, but it does indeed support the biblical record of the Genesis Flood.

Prayer: Lord, with so many conflicting voices and claims in our world, I thank You that Your Word is trustworthy. Help me to see more clearly when the world's claims contradict Your truth, and increase my faith. Amen.

Different Races, One Blood

Acts 17:26
"And hath made of one blood all nations of men for to dwell on all the face of the earth, and hath determined the times before appointed, and the bounds of their habitation;"

How did we get all of the different races on Earth if we are all descended from Adam and Noah and his family?

Today's differing racial and cultural characteristics began at the separation of the world's population into different groups based upon language. This followed the affair at the Tower of Babel. Before this separation, the one population shared the genetic traits of all humanity. In other words, they had a common gene pool. After the separation into groups, each group contained only a part of the former gene pool. This means that each group had lost some of the genetic information. As one generation followed another, genetic information continued to be lost in each group and, in fact, is still being lost to this day. Certain traits in each isolated group became highlighted. For example, the decreasing tendency to produce melanin in the skin gave rise to the Caucasians, the white people of Europe. Other traits like hair type and soft tissue characteristics became unique to various groups that we refer to as "races" today.

Interestingly, interracial marriages produce offspring that recapture some of that genetic information lost by each race represented by their parents.

No matter what race we belong to, we all share the same feelings and desires. Our Creator invites each of us to take to heart the truth that the Son of God became one of us to restore us to our Creator through the forgiveness of sins.

Prayer: Dear Father in heaven. Help me to see that while we are enriched by all of humanity's different cultures, the religious variety of mankind is not pleasing to You, since there is no other Name given to us for salvation than Your Son, Jesus Christ. Use me to spread that Name and what He has done. Amen.

Dinosaurs in History

Genesis 1:25
"And God made the beast of the earth after his kind, and cattle after their kind, and every thing that creepeth upon the earth after his kind: and God saw that it was good."

One of the most common questions Christians ask is: "What about the dinosaurs?" Are dinosaurs proof that the Earth is countless ages old and that life must have evolved?

By themselves, dinosaur fossils do not disprove or contradict any history found in the Bible. Vast ages are not needed to form the Earth's sedimentary rocks and the fossils in them. The flood of Noah formed the rock layers quickly. The fact that fossil bones are found jumbled together in great heaps argues for their deposition during a flood. A global change resulting in a cooler, wetter climate probably caused the extinction of most of the survivors of the flood.

This means that man and dinosaur lived on Earth at the same time. Perhaps we still do. In 1977, a Japanese trawler hauled up the decaying remains of a sea creature that had a long neck and flippers. After taking tissue samples and photographs, they returned the rotting carcass to the sea. The creature was commonly held to be a plesiosaurus, a sea-dwelling dinosaur. Virtually every people on Earth have legends about dragons. Those dragons resemble dinosaurs very closely. If man had never seen these great reptiles, why does everyone seem to know about them?

The great sea-going dinosaurs were most likely created on day five of creation week with the other great sea creatures. The great land dinosaurs would have been created on day six. Dinosaurs don't contradict the Bible; rather, they glorify their Creator.

Prayer: Lord, the dinosaurs were great and mighty creations of Your limitless hand. Help Christians to understand that these magnificent creatures do not call the Bible into doubt. Instead, these mighty creatures glorify You. Amen.

Is There a Third Choice?

Joshua 24:15b
"...but as for me and my house, we will serve the LORD."

Creationists are often asked whether there are only two possible choices in the origins debate, creation or evolution.

Evolution includes a broad range of explanations. Each evolutionary explanation relies on natural forces to explain the creation. Darwinism, neo-Darwinism, pantheistic or cosmic evolution and theistic evolution all rely primarily on naturalistic forces to produce the universe and life. Each denies the Bible's claim that death is a result of sin and, therefore, that Christ's death *paid* for our sin.

Like evolution, creationism is divided into many types. Many different world religions have a Creator who formed the world and its living things by supernatural intervention. Most of them even agree that creation was completed within six days. Most creation stories make the point that the living things were created in their final form. Both orthodox Jews and Muslims accept the biblical account of creation at face value. Most of the world's cultures preserve an account of divine, supernatural creation by a Creator.

In summary, both creation and evolution are broad families of belief offered to explain our origins and are based upon either supernatural or natural forces. Either there is a personal Creator who is intimate with His creation or there is not. The bottom line is that there are only two possible explanations for our origins: the Creator God and His creation or godless evolution.

> ***Prayer: Dear Father, our day and age frowns on those who accept truth as absolute certainty. I know that this can affect my thinking, too. Help me to be bold in my faith and unapologetic about holding to the truth You have revealed in the Bible. In Jesus' Name. Amen.***

The World's Oldest City

Hebrews 11:7
"By faith Noah, being warned of God of things not yet seen as yet, moved with fear, prepared an ark to the saving of his house; by the which he condemned the world, and became heir of the righteousness which is by faith."

When we think of Mount Ararat, we usually think of the search for Noah's Ark. While the Ark has not been definitely located in modern times, the mountain has many other discovered mysteries. What is very likely the world's oldest city sits in ruins on the southern slopes of Mount Ararat. Tradition and evidence seem to support the idea that this was the first city built after the flood.

According to tradition, after the flood, Ham built a city called Naeltamauk. The ruins on Mount Ararat are all that remain of what were clearly buildings and streets. The great stone blocks of the city are cut in the ancient Urartu design. Could Noah have walked the still discernible streets of this city? Could this be the setting for the events recorded in Genesis 9? Might one of these buildings have been Noah's house?

Like so many of the caves and other sites on Mount Ararat, the city is wrapped in mystery and legend. The local people take the Old Testament biblical history very seriously. They seem to have little doubt about the mountain as the landing place of the Ark, Noah's home after the flood, and the burial place of Noah and his wife. Strangely enough, the ruins of the city have never been systematically investigated. Imagine what we might learn, especially if this is indeed the city that Ham built!

We never need doubt that the Bible's accounts of the earliest events of man's history are true. Those who live with the strange monuments on Ararat have no doubt about Genesis.

Prayer: Father, I thank You for revealing to us in Scripture why You sent the flood. I thank You for Your mercy in preserving Noah, his family, and the animals. Let me not fall into the carelessness of those who lived before the flood. Rather, help me to always be ready for the return of my Lord and Savior, Jesus Christ. Amen.

Fossil Hagfish Tells Story

John 7:24
"Judge not according to the appearance, but judge righteous judgment."

Many of God's creatures are beautiful. Others of them, like the appropriately named hagfish, strike us as ugly. Not only does the hagfish look like something out of a science fiction movie, but also it has bad habits.

The hagfish looks like an eel with whiskers. It has no scales. When handled, a slippery slime oozes from its skin, which explains its other name, the slime eel. It lives at the moderately deep levels of the world's oceans. Its single eye is beneath its skin. It has no jaw. It has true teeth as well as barb-like teeth on its tongue. The hagfish eats other fish, sometimes entering them and eating them from the inside out. One wonders why God made such an ugly creature that seems so disgusting.

The oldest evidence of the hagfish is a fossil in ancient rocks. While we can discount the inflated evolutionary years that are used to date these rocks, it does appear that the fossilized hagfish died when the world was still young. Yet, this fossilized hagfish is identical to the modern hagfish. To paraphrase one scientist, no evolution has taken place.

It appears that God has an important job for the ugly and impolite hagfish. The hagfish shows that evolution has not taken place. Beauty, or lack of it, doesn't always witness to our Creator, but the truth unfailingly does!

Prayer: Dear Father in heaven, I thank You for You have revealed Your truth to us in Your Word. Help me to avoid the temptation of judging the worth of something by its appearance, and help me judge things by their witness to the truth, which is found only in Your Word. In Jesus' Name. Amen.

Ref: Andrew Herrman. 1991. "North Sider's Hagfish Story is Really one for the Books." *Chicago Sun-Times*, Nov. 1, pp. 1, 12.

Social Spiders

Philippians 2:3
"Let nothing be done through strife or vainglory; but in lowliness of mind let each esteem other better than themselves."

We can all understand why spiders like to live alone rather than with other spiders. Spiders consider any living creature of eating size appropriate to eat, even if that other creature is a spider. Since spiders are aggressive about eating, they are usually smart enough not to live too close to another spider.

Some Mexican spiders are a bit less self-centered. A few species of spiders build communal webs and share their prey with each other. Some of the female spiders are loving mothers toward their children. These spiders can only build their large communities in tropical areas where prey is plentiful.

Each of this unusual species of Mexican spider builds its own "apartment" within the spider colony. However, it connects its orb-shaped web to the orb webs around it. These communities can be huge, having as many as 7,000 spiders. Each is responsible for defending its own "apartment" and capturing its own prey. But the fact that hundreds or even thousands of orbs are joined gives each spider a better chance of capturing prey. An insect might bounce off a solitary web. However, when the only place the insect has to bounce is into another web, it's inevitable that the insect will be captured.

While selfishness and survival of the fittest sometimes seem to be the way to come out ahead, they produce nothing productive. This offers us a hint of the truth that the Bible clearly tells us. There are general moral absolutes in the creation, including love for others.

Prayer: Dear Lord, I thank You that this is not completely a "dog-eat-dog" world. I ask You to forgive me for those times I love myself more than You and at the expense of others. Give me your peace and help me to love others as I love myself. Amen.

Ref: K. A. Fackelmann. 1982. "The Social Spiders of Mexico." *Science News*, Aug. 7, p. 87.

Stupid Rats and Evolution

Job 39:13a, 14, 16-17
"Gavest thou the goodly wings unto the peacocks?...Which leaveth her eggs in the earth, and warmeth them in dust...She is hardened against her young ones, as though they were not her's: her labor is in vain without fear; Because God hath deprived her of wisdom, neither hath he imparted to her understanding."

Humans have a relatively large brain cortex, and the evolutionary argument has always been a larger brain cortex with more intelligence. In an attempt to prove this theory, researchers at the National Institute of Mental Health gave pregnant rats a drug that reduced the amount of cortex tissue in the next generation. The researchers expected the offspring to be less intelligent, but this was not the case. Instead, the offspring were more aggressive and short-tempered, and there was a tendency towards cannibalism; only ten percent of the young made it to adulthood.

The researchers then concluded that the larger cortex had less to do with intelligence, but it did enable creatures to deal with complex life situations. In other words, the formula for survival of the fittest had shifted from intelligence to the ability to deal with complex situations. This was claimed to be evidence for evolution. It is well known that prefrontal lobotomy (removing the front part of the brain) in human patients does not change intelligence but often leaves the patient antisocial. The truth is, the drugs given to the rats not only reduced the size of the cortex but undoubtedly caused brain damage, leaving them antisocial and unable to handle complex situations or antisocial. In other words, it is not a matter of simple cause and effect but two effects from one cause. Time and again, the theory of evolution has been found to rest upon such flawed experimental work.

Prayer: I thank You, Lord, for my brain and all my senses. Help me to love You with all my heart and all my mind, and all my soul. Help me to employ these to bear witness of You in our dark and confused world. Amen.

Ref: "Evolution of the Cortex." *Science News*, Vol. 122, p. 76.

God's Gift of Pets

Genesis 9:2
"And the fear of you and the dread of you shall be upon every beast of the earth, and upon every fowl of the air, upon all that moveth upon the earth, and upon all the fishes of the sea; into your hand are they delivered."

God did not allow man to eat meat before the Great Flood but gave us permission to do so after the flood. At the same time, God placed the fear of man into the animals so that they would know that they needed to protect themselves.

Those animals we generally refer to as "domesticated", however, remained in a close relationship with man. Even animals we don't often think of as pets can offer benefits to man.

Researchers have found that when autistic children interact with dolphins, the children become more communicative. Researchers say that eye contact with pets is especially valuable. Animals generally find it very important to greet each other after a period of absence. That's why our pets are so eager to greet us when we come home. It's an important ritual for them, and we usually enjoy the warmth of the moment. Researchers have also found that when pets are introduced into a family setting, families generally show more signs of closeness and warmth. There is more healthy playing and less arguing. One of the most surprising findings was that when people give attention to their pets, blood pressure drops. Even gazing into an aquarium lowers blood pressure. People with the highest blood pressure benefit the most from this interaction.

The unconditional love offered by our pets is truly a blessing from God. But our Creator's unconditional love toward us through the forgiveness of sins in Jesus Christ is greater than any other love we can ever experience.

Prayer: Dear Father in heaven, I thank You for the blessings we enjoy through relationships with the animals around us. Most of all, I thank You for Your unconditional love toward me through the forgiveness of sins in Your Son. Amen.

Ref: Joan Arehart-Treichel. 1982. "Pets: The Health Benefits." *Science News*, Vol. 121, Mar. 27, pp. 220-223.

Fast Rocks

Psalm 46:2-3
"Therefore we will not fear, though the earth be removed, and though the mountains be carried into the midst of the sea; Though the waters thereof roar and be troubled, though the mountains shake with the swelling thereof. Selah."

When we think of major geological changes in the Earth, we usually think of events that happen too slowly to notice. Certainly no one would notice major geological shifts in the space of one lifetime, or even the lifetime of a civilization. Yet, if the biblical history is true, major geological changes have taken place within the lifetimes of civilizations.

From 500 A.D. until their invasion by the Incas in 1476, a large and powerful civilization lived in what is now northern Peru. Between 500 and 1200 A.D., they built the largest network of irrigation canals ever constructed in South America. This canal system was built using engineering methods that are considered impressive even by modern standards. The remains of the canal system can still be seen today. Because of geological changes, though, the canals no longer work. As a result, farming is possible on less than one-half of the lands used by the original Indians.

Geologists studying the canal system discovered that portions of the South American continent shifted so rapidly that the Indians had to continually redesign their canal system to keep water in it. The plates on which South America rests changed the slope of the ground so rapidly that the Indians could not keep up with the changing geology.

The plight of the ancient, canal-building Indians of Peru helps us understand that the forces that shaped today's Earth did not need millions of years to do their job.

Prayer: Dear Father in heaven, truly You are the all-powerful God. All of the creation bears witness to You and glorifies You. Grant your people boldness to bear witness to Your forgiving love to us in Christ, even though the world around us may be hostile to Your saving truth. In Jesus' Name. Amen.

Ref: C. Simons. "Why Ancient Canals Went Wrong." *Science News*, Vol. 122, p. 56.

Born to House Hunt

Matthew 6:8
"Be not ye therefore like unto them: for your Father knoweth what things ye have need of, before ye ask him."

We've all heard the old saying, "God helps those who help themselves." It's usually offered to encourage someone to put in more effort. While there are benefits to hard work, it's not exactly accurate to say that God helps those who help themselves.

Hermit crabs are one of the most industrious creatures you will find at the seashore. They never seem completely happy with the shell they have chosen for their home. They seem constantly to try on new shells. Sometimes a hermit crab will try on the new find, return to its old shell, and try on the new shell again just to see which one offers the most comfortable fit.

Obviously there is more to this story than just the fussiness of the hermit crab. God gave hermit crabs the ability to sense calcium – the stuff of which shells are made. Hermit crabs can even sense dissolved calcium from buried shells. Research shows that a hermit crab favors shells that have the most calcium, as long as it's a comfortable fit.

If hermit crabs were unable to find the shells with the most calcium – the strongest shells – they could endanger their lives by adopting unsuitable homes. God gave the hermit crab its special abilities before the hermit crab started using those abilities to help itself. As we review our own accomplishments, we do well to apply the lesson of the hermit crab to ourselves, and learn to use the talents He has given us.

Prayer: I Thank You, dear Father, that You are a generous God Who knows what we need and gives it to us even before we know to ask. I especially thank You that I am saved by grace through faith in Jesus Christ, for I could never meet Your perfect standards on my own. In Jesus' Name. Amen.

Ref: "Hermit Crab goes House-Hunting." 1982. *Science News*, Mar. 6, p. 153.

Frozen Frogs

2 Chronicles 30:22a
"And Hezekiah spake comfortably unto all the Levites that taught the good knowledge of the LORD..."

Four or five centuries ago, people thought that small animals simply froze solid in the winter and thawed out in the spring to resume life. These ideas were later thought of as a foolish fiction of a previous, ignorant age.

Today, science has a growing list of creatures that do indeed freeze in the winter, and thaw and resume life in the spring. Scientists recently added three species of tree frog to that list. As the cool, fall weather sets in, these northern tree frogs usually burrow beneath the forest's leaf litter. If there is little snow, however, the frogs have no protection from freezing. Scientists assumed that the frogs' bodies manufacture antifreeze as do some insects and polar fish.

When scientists collected some tree frogs for laboratory study in the late fall and winter, they received a surprise. Up to 35 percent of the frogs' body fluids froze when the frogs were cooled to several degrees below freezing. When thawed, the frogs returned to normal activities. Scientists discovered that the frogs' bodies produced glycerol. This alcohol acts as an antifreeze. What is more important, it prevents ice crystals from forming in a way that destroys cells.

The created world around us is full of surprises. Because God is God, His unlimited wisdom and power allow Him to create anything in any way that He likes. Because God is Creator, science cannot guess how nature might work. It must investigate to learn the surprising designs God built into the creation.

Prayer: I thank You, Lord, for the blessings of modern science. I ask that You would increase our witness to the truth so that science may be used less and less as a tool to mislead. Prosper all good knowledge to Your glory. Amen.

Ref: "Frogs that can be Frozen." *Science News*, Vol. 121, p. 122.

Biological Balance

Nehemiah 9:6
"Thou, even thou, art LORD alone; thou hast made heaven, the heaven of heavens, with all their host, the earth, and all things that are therein, the seas, and all that is therein, and thou preservest them all; and the host of heaven worshippeth thee."

Did you know that a tree is aerodynamically more sleek than the smoothest jet plane? As the wind blows, the leaves and branches of trees move, allowing the wind to pass. As a result, the tree has less aerodynamic drag than a jet plane.

The creation is full of examples of the Creator's principle that life needs to be resilient to change. This means that, for the sake of survival, life is not rigidly designed but has certain flexibility.

Much is said today about an increase in carbon dioxide in the Earth's atmosphere, and concern is expressed about global warming. But man's production of carbon dioxide is infinitesimal compared to the carbon dioxide from volcanic eruptions. Now science is learning how plants are designed to regulate carbon dioxide in the atmosphere. When carbon dioxide levels increase, many plants increase their absorption of carbon dioxide, at the same time making more oxygen available. With increasing carbon dioxide levels, plants also tend to produce tissue that decays less easily. This ties up the excess carbon for longer periods of time.

Our Creator expects us to take care of His creation. However, He knows that we cannot change conditions on planet Earth that are beyond our control. For this reason, He has built mechanisms into the creation that adjust to stabilize changing conditions. How wise He is in His goodness to us!

Prayer: Dear Father in heaven, I thank You that You have wisely built a creation that can adjust to normal change and help correct imbalances. Help me to take good care of my part of Your creation and remove from me the fears that the world would instill in me. In Jesus' Name. Amen.

Ref: J. Raloff. "Not all Plants will Thrive in a Greenhouse." *Science News,* August 26, pp. 142-143.

Groceries, Ant Style

Proverbs 6:6-8
"Go to the ant, thou sluggard; consider her ways, and be wise: Which having no guide, overseer, or ruler, Provideth her meat in the summer, and gathereth her food in the harvest."

Even a small household must have some organization to keep enough food and supplies on hand for the family. Yet, the thousands of members in an ant colony never run short of supplies, even though there is no central organization and not one shopping list. Modern science is discovering how the ant's astonishing system works.

An ant colony depends on worker ants to bring in water and supplies. In the morning, workers search mainly for drops of dew that can be carried into the nest. Some of the water is shared with nest mates. Some of the water is delivered to the brood chambers where immature ants are still developing. This water is used to soak the ground to keep the humidity inside the chamber high. Food is likewise collected by the workers and returned to the colony to be shared with every member of the colony. Food is most often shared by regurgitation.

To learn how widespread this sharing in the colony is, researchers gave one worker ant sugar water that was radioactively tagged. Within a day, some of the tagged material had been shared with every member of the colony. Within a week, every member of the colony had an equal amount of the tag. This means that through continuous food sharing, every member of the colony always has an equal amount of food. When one worker is hungry, she knows that the entire colony is hungry.

This simple, yet ingenious grocery system designed by the Creator keeps ant colonies supplied through the sparsest of times.

Prayer: I thank You, Father, that there is no end to Your wisdom and Your care for the creation. Help me to remember that You will also care for my needs on Earth as well as eternity, for You have shown me Your love for me in Jesus Christ. Amen.

Ref: E. O. Wilson. 1985. *Discover*, August, pp. 47-49.

Deep Diving Wonders

Genesis 1:21
"And God created great whales, and every living creature that moveth, which the waters brought forth abundantly, after their kind, and every winged fowl after his kind: and God saw that it was good."

The water pressure around a human diver increases as he goes into deeper water. As the pressure increases, his blood is able to hold more dissolved oxygen. Our blood also absorbs the nitrogen in the air around us. If a diver were to move toward the surface too quickly, the nitrogen would start to bubble out of his blood. These bubbles can block the flow of blood to muscles, organs, and even the brain, leading to death. This painful condition is called the bends.

Scientists have wondered why seals don't get the bends. Weddell seals dive to far greater depths than human divers would consider, using even the best equipment. The fact that they are breaking every diving rule in the book means that almost every dive should lead to a fatal case of the bends.

To find the answer to this mystery, scientists outfitted four seals with scientific backpacks. These allowed scientists to record the seals' heart rates, sample blood, and record the depths of their dives. The deeper the seals went, the more nitrogen accumulated in their blood. Just before the nitrogen reached a dangerous point, it leveled off. Scientists say that the tiny sacks in the lungs that absorb oxygen and nitrogen shut down. Then the seals' heart, liver, and blubber begin to absorb the nitrogen from the blood. The air exchange sacks in the lungs reactivate as the seal ascends to the surface.

Surely the amazing biology that allows the seal to make his living deep in the ocean could only have been designed by our Creator God.

Prayer: Dear Father, as science learns more about the wonderful things You have created, I am moved anew to glorify You. I pray that our growing knowledge of what You have made will lead others to do the same, especially those who don't know You or the forgiveness of sins in Your Son. Amen.

Ref: "Why Seals Don't get the Bends." *Discover*, October. 1985, pp. 10, 12.

Your Brain's Produce Section

Job 38:36
"Who hath put wisdom in the inward parts? or who hath given understanding to the heart?"

Your grocery store has a produce section. Recent discoveries suggest that your brain has a produce section, too.

We use many different kinds of filing systems to keep track of different kinds of information. Your medical records are stored using a filing system that is quite different from the system used to keep track of your subscription to your favorite magazine. While we can think of many different kinds of filing systems, no one has ever discovered the filing system used by the human brain. We simply don't understand how it keeps track of things.

Researchers are wondering whether a stroke victim might offer a clue to how the brain arranges information. One man suffered damage to the left frontal lobe of his brain. As is common with this kind of damage, he had difficulty naming things. This patient's difficulty, though, was most unusual. He could identify just about anything. He could look at pictures, memorize lists of words, and do as well as anyone else, except for one class of things. He could not identify the picture of a fruit or vegetable. He couldn't remember any fruits or vegetables that he memorized out of a word list. One scientist said that it was just as if he had lost the page in his brain that had fruits and vegetables written on it.

We may never be able to understand the human brain fully. But clearly an organ this wondrous and complex *had* to be designed and created by an even greater mind. That Creator desires a day-by-day relationship with you through His Son, Jesus Christ.

Prayer: Dear Lord, when I see how wondrous the human brain is, I am humbled, and I know that I could never completely appreciate You. Do not let this prevent me from having a closer relationship with You through faith in Your sacrifice for me on the cross. Amen.

Ref: "Food for Thought: Does the Brain have a Produce Section?" *Discover*, October 1985, p. 10.

Survival of the Generous

Romans 8:22
"For we know that the whole creation groaneth and travaileth in pain together until now."

On television's nature programs, we often hear that the principle of survival of the fittest rules in nature. The most aggressive flourish at the expense of those who don't kill or steal quite as quickly.

This principle has even been applied to humans. Some would say that in business, it's the most aggressive and the most merciless to enemies who prosper. Is this a fair picture of nature or man?

African wild dogs are able to bring down their prey because they hunt in cooperative packs. Once they have eaten, they then take food to the rest of the pack. There, even injured members of the pack receive their share of the food. Likewise, adult male chimpanzees hunt for the entire community, and social insects cooperate in the same way.

The truth is that without cooperation, life would be brutal, if not impossible. If nature's principle were indeed survival of the fittest, many creatures alive today would have become extinct long ago. Cooperation aids survival. Violence, selfishness, and raw aggressiveness can be found both among man and animals. But where they exist, the quality of life for man and animal is decreased. Survival of the fittest isn't the ethic that produced life. The ethic of survival of the fittest is evidence that man's sin has made the creation something other than what God intended. It's evidence that we need the forgiveness of sins that God has provided for us in our Lord and Savior, Jesus Christ.

> **Prayer: Deliver us, Lord, from those who seek to make life more brutal, and fill Your people with peace among the aggressive. Help my generosity and love reflect Your love and generosity to those around me so that they may be invited to Your forgiving grace. Amen.**

Ref: E. O. Wilson. 1985. *Discover*, August, p. 48.

Bird of Paradise

Genesis 2:19
"And out of the ground the LORD God formed every beast of the field, and every fowl of the air; and brought them unto Adam to see what he would call them: and whatsoever Adam called every living creature, that was the name thereof."

The bird of paradise is among the strangest and most beautiful birds in the world. It is so unusual that evolutionary scientists are having difficulty explaining how such creatures could have evolved.

It was not until 1824 that the first European saw a bird of paradise. Before that time, the bird was known in Europe only by the skins and feathers that natives had sold to European feather dealers. Since the skins did not include the feet, Europeans thought that the bird of paradise remained in flight its entire life. Its Latin name even means "of paradise, without feet."

The male bird of paradise is probably one of the most beautiful birds in the world. Its body is a dark maroon, with bright green feathers. The 42 different species have an assortment of bright yellow plumes, tufts, and other decorations. At mating time, about all one can make of the bird is a wildly screaming yellow tuft that bounces from branch to branch, zigzags in the air, and hangs upside down from branches. Science has always thought that this odd behavior, and the bright plumage, served to attract female attention. This was the justification for these traits to evolve, according to evolutionists. More recent research shows that females do not select males – the males select the females. Evolutionists are left with no way to explain the bird of paradise.

The beauty and odd behavior of the bird of paradise are not hard to explain for those of us who know that our Creator has unlimited creativity and inventiveness.

Prayer: Father in heaven, I thank You for making the wonderful bird of paradise. I especially thank You for the way in which this creature glorifies You among men. Help me to glorify You as You have created me to do. Amen.

Ref: Laura Tangley. 1982. "Sexual Selection's Strangest Inventions." *Science News*, Vol. 122. September 4, pp. 152-170.

A Case of Bad Dates

John 10:35b
"...and the scripture cannot be broken..."

Scientists who believe in creation have always said that evolutionary methods for dating the ages of rocks give ages that are too old. There are many examples of how those who believe in evolutionary long ages exaggerate the abilities of their dating methods. Scientists at the Institute for Creation Research recently published another example of how evolutionary dating methods give wildly exaggerated ages.

The dating methods that give the most ancient ages for rocks work on rock that was once lava. In the Grand Canyon of the western United States we find two rock formations that were once lava. One is intruded deep in the canyon. It is dated by evolutionists as among the oldest rocks in the canyon – over a billion years old.

There is a second lava flow, universally recognized as the youngest rocks, that flowed from the top of the rim of the canyon all the way to the river at the bottom of the canyon. This flow is dated as merely thousands of years. However, scientists from the Institute for Creation Research had samples from this flow dated at three laboratories, using the same method used to date the older flow. While the rock should have been too young to produce a date, it was actually dated as being older than the rock at the bottom of the canyon!

This is still more scientific evidence that those huge evolutionary ages are not well supported by science. There are no good scientific challenges to the truth of the Bible.

Prayer: I thank You, Lord, that among all the words in the world, Your Word is trustworthy. Bless those who are working to reveal the lies that unbelievers would heap upon Your truth, and increase my commitment to study it every day. Amen.

Ref: Steven A. Austin, Ph.D. 1992. "Excessively Old "Ages" for Grand Canyon Lava Flows." *Impact*, February, p. I-iv.

One-Way Only

Acts 4:12
"Neither is there salvation in any other: for there is none other name under heaven given among men, whereby we must be saved."

Not a week goes by that we don't hear of some plant or animal threatened with extinction. Is such a rapid rate of extinction natural, or is it caused by the activities of modern technological man?

It is popularly thought that the extinctions of various species on the Hawaiian Islands resulted when Western civilization began to move there. However, studies of bird fossils on the islands show that many more species became extinct before we arrived. In the last 1,500 years, one-third of all the species on the islands have become extinct.

Unquestionably, creatures are becoming extinct while no new creatures are evolving. Nor can there be any question that these extinctions are natural and have little to do with the growth of modern technological man on islands or anywhere else. This shows that we could not have gotten the millions of different life forms we know on Earth today from evolution since creatures naturally become extinct faster than they are supposed to have evolved.

This is a warning to us that the creation is moving one direction only. That direction is downward toward death. Realizing this inescapable fact, we need to see another one-way sign. It's found in the Bible where Peter tells us that there is no other name given among men whereby we must be saved. That Name is Jesus Christ. Only He has conquered death – and He has done so on our behalf, rising back to life to prove His power to forgive sin.

Prayer: *I thank You, dear Father in heaven, that You sent Your Son, Jesus Christ, to rescue us from this decaying world. As my sins are forgiven through Him, help me to put away sin in my life and have Your peace. In Jesus' Name. Amen.*

Ref: C. Simon. 1982. "Revising the Record of Island Birds." *Science News*, August 14, p. 103.

Take Heart! Be Bold!

2 Corinthians 3:12
"Seeing then that we have such hope, we use great plainness of speech:"

Do you believe in a recent, special creation by God? Those of us who do are often ridiculed as being on the fringe. The suggestion is that only a small minority of people believe in a recent, special creation by God. A Gallup Poll shows that those of us who believe a recent, special creation actually outnumber those who believe in atheistic evolution or even those who believe that evolution was directed by God.

When people believe that they are in the majority they usually feel their beliefs are justified and try to show that those who disagree with them are in a minority.

Those who have felt this way can now take heart. A Gallup Poll released in November of 1991 has shown that only 9 percent of the U.S. population believes in atheistic evolution. Another 40 percent believe in some form of God-directed evolution. However, the fastest growing segment is also the largest – those who believe that God recently created all things as described in Genesis!

Your belief in creation may be ridiculed by some people as a fringe view. But the common-sense view of special creation by God is gaining new followers every day. So be bold in your witness of the truth that God created the heavens and the earth, not billions of years ago, but only a few thousand.

Prayer: Dear Lord, make me bold and unafraid to speak Your truth to others. Help them to understand that because You made us, you hold us responsible for our actions. And because You love us, You earned forgiveness for us on the cross. Amen.

Ref: John D. Morris. 1992. "Do Americans Believe in Creation?" *Back to Genesis*, no. 33, Feb, p. d.

Flying Through Water

Lamentations 3:39-41
"Wherefore doth a living man complain, a man for the punishment of his sins? Let us search and try our ways, and turn again to the LORD. Let us lift up our heart with our hands unto God in the heavens."

It's not at all true that penguins are flightless birds. They are perfectly designed to fly. When a penguin goes flying in search of its food, however, it doesn't search in the air, but in the water.

The king penguin is one of the world's greatest flyers in water. Scientists have attached scientific recording devices to the backs of king penguins. To their surprise, they learned that the bird dives deeper than 600 feet in search of its prey. This is deeper than any other penguin except the emperor penguin. It regularly dives to depths greater than 300 feet. Scientists also studied the penguins' metabolism. They were impressed by the fact that dives to 100 feet take very little effort from the penguins. Some birds averaged over 300 dives to various depths each day! This is especially impressive when you realize that the bird is holding its breath while making these dives.

The king penguin must reach such great depths because its primary diet, squid, are found at these depths. Scientists estimated that between 50 to 90 squid must be caught by the penguins on each trip to the sea. There must be enough squid to provide energy, not only for the diving parent, but also for the chick back in the nest.

We should not think of the penguin as a bird that lost the ability to fly. The king penguin's wings are not designed for flight in the air. Instead, they are perfectly designed for flight in water. That's proven by the fact that they can reach such great depths and are so effective in catching squid. Even penguins glorify their Creator!

Prayer: *I thank and praise You, Lord, that nothing in the creation fails to glorify You. Forgive me for the times I have complained about something, and fill me with thanksgiving and the urge to glorify You before men. Amen.*

Ref: J. A. Miller. 1982. "Deep Dives at Sea with King Penguins." *Science News*, August 21, p. 119.

Those Dropping Science Scores

Psalm 14:1
"The fool hath said in his heart, There is no God. They are corrupt, they have done abominable works, there is none that doeth good."

We've been hearing a lot about how science scores are dropping among American high school students. Many creationists believe that the dogmatic teaching of evolution is one important reason that science scores are dropping.

Today, evolution creeps into many of the subjects taught in America's high schools. Both biology and history classes have units populated with ape-like creatures supposedly on their way to becoming human beings. Students are taught how religion evolved when ancient man invented God. Classes on ecology and natural history dwell on supposed evolutionary ages and developments.

Consider these claims from the viewpoint of a student. Evolution claims that non-living matter once formed the first living creature *all by itself.* Students are taught that a small rodent turned into a horse. They are told in all seriousness about how fish once started walking around on land, and so grew lungs. Often they are not allowed to question these outrageous claims. From the students' viewpoint, science departs from reality and common sense. It becomes confusing, and the student's choice is either to memorize the stuff, pass the examination and have done with it ... or drop out.

When high school students hear a critical analysis of evolution and are shown the evidence supporting creation, they quickly become interested. They never realized that science could make sense or that true science does not call their belief in God into question.

Prayer: Dear Father in heaven, I ask You to support all our Christian young people as they are educated. Prosper all efforts to give them a sound education that glorifies You, and help them be examples to the other students. In Jesus' Name. Amen.

Ref: John D. Morris. 1992. "Do Americans Believe in Creation?" *Back to Genesis*, no. 33, Feb., p. d.

Where Did Learning Come From?

Psalm 111:10
"The fear of the LORD is the beginning of wisdom: a good understanding have all they that do his commandments: his praise endureth for ever."

It appears that all animals have the ability to learn. This is because they need information about their surroundings that will help them predict other things about their surroundings. For example, honeybees know which types of flowers offer pollen at any given time. They don't waste their time checking out flowers that have nothing to offer.

Scientists have been amazed at how a wide range of animals shows two different types of learning. Even a garden slug can learn to avoid its favorite food through simple conditioning. Many animals also learn from changes in their environment. Animals learn to expect different results if their schedule is changed, for example. Birds learn the song their species sings by hearing their elders sing. They also learn regional variants of their species' songs.

Scientists who believe in evolution are struggling to explain how the ability to learn could have evolved. They are struck by the fact that learning follows the same principles in pigeons, bees, and slugs. At the same time, they admit that they don't know how learning works. We need to ask: How could they know learning evolved if they don't even know how it works? The fact that they are sure the ability to learn could have evolved, even though they don't know how, shows that evolution is a faith and not science.

We learn a better faith when we go to the Book that was written by the Author of all learning. The Bible tells us about His will and explains that He offers us the forgiveness of sins in Jesus Christ.

Prayer: Lord God, You are the Author and Source of all knowledge. Help me to use the knowledge I have to Your glory, and help me to constantly seek to learn more about You in Your holy Word, the Bible. In Jesus' Name. Amen.

Ref: Julie Ann Miller. 1983. "Lessons from the Lab." *Science News*, Vol. 124, December 17, pp. 394-396.

Deception of the Bola

2 Thessalonians 2:9-10
"Even him, whose coming is after the working of Satan with all power and signs and lying wonders, And with all deceivableness of unrighteousness in them that perish; because they received not the love of the truth, that they might be saved."

The bola spider is said to have evolved its amazing abilities all by itself. Why, then, is it smarter and gifted with more abilities than mankind?

The bola spider's favorite food is moths – but it doesn't weave a web. The bola spider produces the same scent produced by a female moth. Male moths are attracted to this scent, and the spider waits for the inevitable appearance of a male moth.

As the male moth nears, the bola senses the vibrations from its wings. At just the right moment, the spider casts out its bola. The bola, after which the spider is named, is a special spider-silk thread with a sticky glob on the end. Scientists who have studied this glue-glob say that modern chemistry could not equal it. The silken line is so elastic that it can stretch up to six times its original length. This allows the spider to conserve its energy while the moth struggles. Once reeled in, the bola wraps the moth in silk for storage until it's ready to eat the moth. The bola's method works so well that it is not unusual for one spider to catch up to eight moths a night.

Normally a species is unable to detect the mating scents of other species. How could the bola spider devise its plan? Once the plan was invented, how could the spider learn how to duplicate the female moth's chemistry? How could the bola spider be more clever than our best chemists? Of course, the only answer to these questions is that the bola spider was designed and created by God, Who invented chemistry in the first place.

__Prayer: Dear Lord, the devil seeks to deceive and mislead me into both earthly and eternal destruction. What's worse is that the world and my own flesh try to help him. Therefore, I seek refuge in You. Help me to use Your Word as a light that dispels his deception. Amen.__

Ref: D. Franklin. 1983. "Prey Caught with a Smell and a Stinker." *Science News*, Dec. 17, p. 389.

Baby Talk

Luke 9:48
"And said unto them, Whosoever shall receive this child in my name receiveth me: and whosoever shall receive me receiveth him that sent me: for he that is least among you all, the same shall be great."

The evidence continues to mount that human beings, unlike any other creature on Earth, are born with language ability. We even begin learning the unique characteristics of our own languages at a young age.

Research shows that children as young as two months old have already learned to distinguish characteristic vowel and consonant sounds in whatever language those around the children regularly speak. Researchers rewarded infants who responded with a head turn in the direction of the test sound. Vowel and consonant sounds distinctive to English and Swedish were used. American infants responded to English sounds two-thirds of the time. Swedish infants ignored these sounds as gibberish. Swedish infants responded to the Swedish test sounds two-thirds of the time. American infants ignored these as gibberish.

Researchers say that this shows that the children have already begun classifying the elements of language that they hear every day. This is the first step in organizing these sounds into meaningful words. Other research shows that by six months of age infants already understand the basic emotional tones of language. Learning to speak takes longer.

The human ability to learn language appears to be pre-programmed into the infant brain. God created human beings for a relationship with Himself. Because He loves even the least among us, He gave human infants built-in language skills. He couldn't wait to begin communicating with each new human being born!

Prayer: You, Lord, are the Creating Word made flesh for our salvation. I thank You for the gift of language. Help my language praise and honor You. Help me to tell others about Your forgiving love for us. In Jesus' Name. Amen.

Ref: "Baby's first Phonemes." *Science News*, Feb. 8, 1992, p. 91."The Rules of Baby Talk." *Science News*, Vol. 122, p. 216.

The Amazing Mexican Free-Tailed Bat

Job 5:8-9
"I would seek unto God, and unto God would I commit my cause: Which doeth great things and unsearchable, marvellous things without number:"

The largest community of mammals in one space spends the summer in Bracken Cave, outside San Antonio, Texas. Here, 20 million female Mexican free-tailed bats raise their twenty million pups, gulping down 150 tons of insects every night!

The bats winter in Mexico and mate in the spring. Then, for some reason unknown to scientists, the females head to Texas. While they make the difficult migration, the females store the male sperm in suspended animation. Once settled in at Bracken Cave, they become pregnant. Four months later, each bat has a single pup. Although the cave's one-room nursery has 20 million noisy pups, a mother bat can find her own youngster in as little as twelve seconds!

Each night, each female leaves the cave and flies as far as 60 miles, consuming as much as her total body weight in insects. After about five hours, she returns to the cave to nurse her pup. The bat's normal diet does not provide enough fat for the mother bat's rich milk. But the nursing period also falls at the same time that an ant, that is a rich source of fat, grows wings. These flying ants are available, in the bats' airspace, at just the right time!

Someone has remarked that God must like bats a lot. He made more species of bats than any other mammal. And He has designed nature to provide for all the bats' needs.

Prayer: Lord God, heavenly Father, I am filled with awe when I look at what You have created. I ask that Your creating hand would be more evident to all who study the creation. Help us Christians to better live and witness the gospel of the forgiveness of sins in Jesus Christ so that more scientists may come to know You. Amen.

Ref: Carol Ezzell. 1992. "Cave Creatures." *Science News*, Vol. 141, Feb. 8, pp. 88-90.

Benefits from Rejecting Evolution

Daniel 1:20
"And in all matters of wisdom and understanding, that the king enquired of them, he found them ten times better than all the magicians and astrologers that were in all his realm."

He lived in the last half of the nineteenth century. He was a strong believer in the Creator God, and that made him very unpopular in some circles. As he pursued his scientific studies in Paris, he became a thorn in the side of his contemporaries, the followers of Charles Darwin. They argued that life arose spontaneously long ago from non-living material, but Louis Pasteur believed that life was created by God. He decided to apply his scientific expertise to Darwin's claims.

His careful experiments, conducted 150 years ago, showed that life could only come from pre-existing life. He effectively put out of business the claim that life could arise from non-living chemicals. His research also showed that fermentation was caused by tiny yeast cells. This led him to discover that many diseases were caused by living organisms too small to see. As a result of his scientific work disproving important evolutionary claims, Louis Pasteur developed vaccines to combat diseases like rabies, diphtheria, and anthrax.

Today, millions of people enjoy better health and safer food because Pasteur sought to show that true science does not disprove the Bible. We can thank God that Pasteur did not sit back and quietly accept atheistic, materialistic claims of his day. He once commented that the more he learned in science, the simpler his faith in the plain words of Holy Scripture became.

Prayer: I thank You, Lord, for the blessings You have given to us through the service of those scientists who faithfully believe Your Word and apply its truths to their research. Grant us more of these workers and the witness and blessings they provide. In Jesus' Name. Amen.

"Cave Men" or Men

Matthew 25:40
"And the King shall answer and say unto them, Verily I say unto you, Inasmuch as ye have done it unto one of the least of these my brethren, ye have done it unto me."

There are numerous sites where evidence that people once lived in caves has been studied. The idea that the first primitive human beings, on their way from becoming apes to humans, lived in caves is ancient. Greek and Roman evolutionists wrote about it over 2,000 years ago. Were people who lived in caves truly less human than we are today?

When we think of cave men, we usually think of creatures that are less than human. Evolutionary belief has had its effect even on minds that accept the truth of the Bible's account of creation. Several years ago, a group of so-called "primitive" cave people called the Tasaday were reportedly discovered in the Philippines. Dozens of scientific papers and articles were written about them, showing how they supposedly represented ancient man. Then it was discovered that the entire tribe was a hoax.

Recently, Indonesian Christians made contact with another cave-dwelling people. These people, the Basap, live in a remote area of Indonesia. The Christians presented the gospel to the chief and left. Nine months later, they returned to find the fruit of their labors. One hundred members of the tribe said that they believed that Christ is their Savior. Baptisms were carried out with the chief's blessing.

The lesson is that from the very first individual, human beings have been fully and completely human. Each has been a creation of God, loved and sought by Him. And God has a special place in His heart for those we often consider the very least among us.

Prayer: I thank You, dear heavenly Father, that You have loved me, not based on my own goodness or worth, but based on Your mercy and forgiveness in Christ. Help me to love others, even if I see nothing lovable in them. For Jesus' sake. Amen.

Ref: "Harvest Among Cave-dwelling Basaps." *The Church Around the World*, Feb. 1992.

Fossil Drives Evolutionists Batty

Psalm 9:1-2
"I will praise thee, O LORD, with my whole heart; I will shew forth all thy marvellous works. I will be glad and rejoice in thee: I will sing praise to thy name, O thou most High."

I'm sure everyone has noticed that whenever one of those interesting nature programs on television show an especially clever design in nature, the program is always quick to explain how this came about by evolution.

One good example is the unusual hearing ability of the moth. Night-flying moths have hearing organs that are able to hear the ultrasonic signals used by bats. The bats use these signals to navigate and to find food, which includes the moths. But evolutionists never want anyone to think about a Creator Who might have given the moths the ability to hear one of their main predators. So they quickly explain that as moths evolved, they developed the ability to hear the bats in response to bat attacks.

A few years ago, a fossilized egg from a night flying moth was discovered. Evolutionists said that the egg was half again as old as any evidence of the first bat. Yet, the egg clearly belongs to a moth species that can hear bats. Evolutionary scientists said that this discovery was most puzzling. They said that evolution would have no reason to evolve this species of moth before bats existed.

Those of us who believe the Word of our Creator in the Bible realize that all living things were made within a few days of each other. We see around us how each was designed in a way that keeps in mind the design found in other creatures. Moths have always been moths, and bats have always been bats. There's nothing puzzling about it at all.

Prayer: Lord, I pray that You would help those misled by evolution to see their foolishness and repent. I ask that You would confound the efforts of those who seek to mislead Your people with stories of evolution and keep us under Your protection. Amen.

Ref: "Ears, Bats, and Early Moths." *Science News*, Feb. 12, 1983, p. 107.

The Search for Extraterrestrial Intelligence

Romans 8:20-21
"For the creature was made subject to vanity, not willingly, but by reason of him who hath subjected the same in hope, Because the creature itself also shall be delivered from the bondage of corruption into the glorious liberty of the children of God."

A century before the birth of Christ, a Roman named Lucretius suggested that since life evolved on Earth, one might expect that there are other planets like Earth where life also evolved.

Modern scientists who believe in evolution also feel that if life could be found on other planets, their arguments for evolution would be stronger. They have persuaded the United States Congress to approve millions of dollars to search for radio signals that might come from other planets. The centerpiece of their effort is a radio receiver linked to a computer. The system can simultaneously listen to over eight million channels for any sign of space communication.

A few scientists are critical of this effort. For any theory to be regarded as scientific, it must be capable of disproof, not proof. Almost anything can be proven by selection of evidence. The critics point out that, given the size of the universe, we could spend billions of dollars and never disprove that there is life on other planets. In this way, it leaves the situation open-ended, and it can always be argued without proof that there may be life.

The Bible offers no direct mention of life on other planets. It does tell us that the entire creation was subjected to the consequences of man's sin here on Earth. Likewise, Christ's atonement for sin on Calvary delivered the entire creation from the bondage of sin. This suggests that our Earth is the only place in the universe where morally responsible beings live.

Prayer: Lord, You have created a huge universe with wonders far beyond our understanding. I thank You that when we ruined the creation, You loved us so much that You gave Your life that we might again be restored to fellowship with God through the forgiveness of sins. Amen.

Ref: D. E. Thomsen. "NASA to Help Listen to E. T.'s Call." *Science News*, Vol. 123, p. 52.

Coma Healing?

Ezekiel 34:4
"The diseased have ye not strengthened, neither have ye healed that which was sick, neither have ye bound up that which was broken, neither have ye brought again that which was driven away, neither have ye sought that which was lost; but with force and cruelty have ye ruled them."

Pastors or clergymen are sometimes called upon to stand with saddened families by the bedside of a loved one who is in a coma. The coma is usually the result of a serious injury, and some have wondered if it is really the brain's strategy for healing itself. Now both experience and science are beginning to suggest that a coma is indeed part of the brain's strategy to heal itself.

Research shows that when a person is in a coma, the brain is actively working to suppress normal conscious activity. Medical researchers say that this suggests that a coma is not the direct result of an injury but a condition that is under the control of the brain. Scans of brain activity show unusual activity in the brain centers controlling muscle activity.

These findings suggest that people who spend months or even years in a coma or vegetative state may be actually in the process of returning to health. This would explain the common stories about someone waking after months, or longer, and rapidly returning to normal.

Modern science admits that it remains largely ignorant about the human brain and how it works. In some cases, people who may have been undergoing a normal healing process have been killed by those who support euthanasia. The Bible has the best advice all the time. Only God can give life, and only He knows when it is over.

Prayer: I pray, dear Father, that we may soon see a change in public sentiment away from so-called "mercy killing" back to Your standards. Help science to better understand the reasons why only You can make life and only You know when it is ended. In Jesus' Name. Amen.

Ref: J. A. Miller. "Coma as a Legitimate Brain Activity." *Science News,* Vol. 122, p. 310. Callista Gould. 1992. "Two Real Life "Awakenings" Challenge PVs Diagnosis." *National Right to Life News*, Jan., p. 34.

The Armed and Dangerous Fungus

Psalm 8:1
"O LORD our Lord, how excellent is thy name in all the earth! who hast set thy glory above the heavens."

It's an armed and dangerous fungus with a hair trigger. The fungus known as *haptoglossa* uses high-velocity projectiles fired from automatic cannons built into its body. Thankfully, the fungus's impressive weaponry is aimed at other microscopic creatures.

Haptoglossa lives in ponds and wet soil. Its prey is the microscopic rotifer. Rotifers are so tiny that they can as easily swim in pond water or the water between the grains that saturates moist soil. When the rotifer brushes against one of *haptoglossa's* cannon cells, the cannon fires a missile of cellular material into the rotifer. It's thought that the cannon's fire power comes from high-pressure fluid at the base of the cell. The fired missile punches a hole through the rotifer's protective covering, preparing the way for a second attack wave.

Next, the fungus extends a hypodermic tube into the rotifer. This tube delivers *haptoglossa's* single-celled infection unit into the rotifer. This cell begins to multiply inside the rotifer until the entire rotifer is nothing but fungus cells.

No matter how small and seemingly unimportant to us, everything God made reflects His creativity and excellent workmanship. Even microscopic creatures are filled with devices and inventions that move us to marvel at the Creator's care for each life form He has made. As Scripture says, "How excellent are His works in all the Earth!"

Prayer: I thank You, Lord, that everywhere I look in the creation I am filled with wonder at Your excellent workmanship. Help me to likewise put excellent workmanship in all that I do so that others may be led to glorify You by my work. Amen.

Ref: J. A. Treichel. 1983. "'Peacekeeper' Fungus: Rotifers Beware." *Science News*, Jan. 8, p. 23.

The Wrong Pattern

Romans 1:21
"Because that, when they knew God, they glorified him not as God, neither were thankful; but became vain in their imaginations, and their foolish heart was darkened."

Does the belief in evolution adversely affect the medical and the social sciences? It sure does.

One example of evolution's negative effect in helping people is a technique called "patterning". "Patterning" is a common treatment for children with brain damage and language disorders.

The theory behind patterning says that the human nervous system must develop by passing through all its earlier evolutionary stages before it works properly. As a result, children with these problems are put through a series of movement exercises that are supposed to mimic the evolutionary development of the nervous system. First the child crawls, then creeps, and finally is supposed to walk around like an ape. This is supposed to help the brain get organized by re-patterning its evolutionary development.

Does it work? There's no reason for it to work if evolution never took place. In the early 1980s, American pediatricians did an in-depth study of patterning. They said that they could find no evidence that patterning helps the child at all. They also warned that patterning could even be harmful to children because the program places great demands on the children and their families. It can create unnecessary stress in families and lead to neglect.

Scripture tells us that those who reject the Creator sink into ever deeper levels of ignorance, all the while claiming that they are wise. Here's an example of where modern science has illustrated this biblical truth.

Prayer: *I thank You, dear Father, for the light of Your Truth in Your Word. Deliver me from ungratefulness and ignorance and help me to show forth the excellence of Your truth. In Jesus' Name. Amen.*

Ref: W. Herbert. 1982. "Treatment for Brain Damage Under Fire." *Science News*, Dec. 4, p. 357.

God's Agriculture and the Stink Bug

Genesis 3:18
"Thorns also and thistles shall it bring forth to thee; and thou shalt eat the herb of the field;"

The world originally created by God was perfect in every respect. However, among the consequences of sin mentioned in Genesis 3 are thorns and thistles, which make man's agriculture more difficult. We could certainly classify insect pests in the same category as noxious weeds.

God did not leave us without earthly solutions to these problems, however. Modern agriculture has relied largely on powerful herbicides and insecticides to protect crops from pests. At the same time, God's design of nature has always included more natural, less toxic ways of controlling pests. Modern science is learning how to use the same methods God uses to control insects.

The spined soldier bug, also known as the stink bug, is one of God's natural controllers of insect pests. The stink bug eats over a hundred different kinds of insect pests, including some of the worst, such as the cotton bollworm and the gypsy moth. Researchers have found that spraying a crop with natural attractant, produced by the stink bug, serves as a call to dinner. Stink bugs naturally move in and begin devouring harmful pests without damaging the crop. Since the stink bug attractant is a naturally-produced chemical, it does no harm in nature before it breaks down.

Our Creator's solutions to problems are always better than ours. You can learn more about His solutions to your problems, including the problem of sin, in the Bible.

Prayer: Dear Lord, increase my appreciation for Your solutions for problems in my life so that I yearn more strongly to learn Your Word and apply it to my life. Where I fail, and where, with Your help I succeed, help me remember that I am not saved by my own efforts, but by the forgiveness of my sins through You. Amen.

Ref: J. A. Miller. 1982. "Chemical Mobilizes Soldier Bug." *Science News*, Dec. 11, p. 373.

Educated Slugs

Psalm 105:2
"Sing unto him, sing psalms unto him: talk ye of all his wondrous works."

So complex is the human brain that even some of today's most visionary scientists have commented that they doubt science will ever fully understand how the brain works. To better understand the brain, scientists have been studying the brains of so-called simple creatures like the garden slug.

But how similar is the brain of a garden slug to the human brain? Researchers have been amazed at the unexpected similarities and abilities shared by both the garden slug and man. Researchers have found that garden slugs can be trained, using unpleasant flavorings, to avoid their favorite food, potatoes. Eventually the slugs will avoid potatoes that have not been made to taste bad. Slugs, and even garden snails, can learn a sequence of events.

How much more simple are the slugs' brains than ours? To their surprise, researchers discovered that the slugs' brains use some of the same chemical methods to learn and store information as do mammals. In other words, the slugs' brains don't use a simpler method to learn and remember than do a cat or a dog's. Mammal brains and slug brains are based on the same design.

This finding came as a surprise and erases evolutionary distinctions between simpler and supposedly more-evolved creatures. It reveals the Creator's pattern. He used the same design when He equipped the slug's brain to learn as He used when He made a dog able to learn. Life is evidence of a perfect Creator, not accidental improvements!

Prayer: Dear Father in heaven, I thank You for the ability to learn. I ask that as we learn more about Your creative work, more people would learn to see Your Hand and seek peace with You through the forgiveness of their sins in Jesus Christ. Amen.

Ref: Robert Pollie. 1983. "The Educated Nervous System." *Science News*, Vol. 123. Jan. 29, pp. 74-75.

Your Fail Safe Heart

Psalm 75:1
"Unto thee, 0 God, do we give thanks, unto thee do we give thanks: for that thy name is near thy wondrous works declare."

Your heart will beat some 100,000 times today. That's over 36 million heartbeats a year and over 2.5 billion times in a 70-year life span.

A healthy heart ticks along, producing beat after beat, whether we are awake or asleep. If we become more active, the heart increases its beating to meet the increasing needs of the rest of our body. Doctors tell us that it's amazing how few of these beats are faulty. They say that it's perfectly normal for even a healthy heart to produce an occasional irregular heart beat. Sometimes an irregular heartbeat is noticeable, but most often it's not. Doctors say that when your heart seems to skip a beat, it has really only beat prematurely. The premature beat leaves a pause before the next regular beat, making it feel as if your heart skipped a beat.

The clockwork precision of the heart's continuing beats is controlled by a built-in pacemaker. The pacemaker, called the sinus node, is a group of cells in the heart's upper right chamber. Research has shown that every cell in the heart is able to send the electrical signal needed to produce a heart beat if the sinus node fails. Using highly complex computer models, medical researchers have only begun to understand the electrical action within the heart.

The human heart is much more than a pump, as once believed. It is also a computer and regulator. Every beat of this wondrously designed biological machine glorifies the Creator Who made it!

Prayer: I thank You, Lord, that You have so wonderfully designed the heart. I join my mind with my voice in praising and thanking You with my whole heart. Remind me of Your often silent but important blessings when my thankfulness grows cold. Amen.

Ref: "Offbeat." *Fairview Healthwise*. P. 7. I. Peterson. 1983. "A Computer's Heart: Simulating the Heart's Electrical System." *Science News*, Mar. 19, p. 183.

Flipperpithicus

Psalm 92:5-6
"O LORD, how great are thy works! and thy thoughts are very deep."

Creation scientists have often complained that evolutionists seem unscientifically eager to turn any ancient fossil into proof of human evolution. In the past, bones from pigs, monkeys, alligators, horses and even an elephant have been "reconstructed" into missing links between humans and ape-like creatures.

While these examples are all old, evolutionists continue to be overly enthusiastic about using animal bones to make human-like creatures. A more recent example caused the respected magazine *Science News* to note that even some evolutionists have begun to discuss "anthropologists' over-zealous pursuit of human ancestry."

In 1979, a bone discovered in northern Africa was described as a hominoid clavicle. In plain English, the bone was said to be the collar-bone of a creature that was an ape becoming man. Its discoverer said that the bone indicated that its owner had possibly even walked upright, like modern humans. As in most of these claims, its discoverer said that it was the oldest evidence yet for the evolution of modern man. Others studied the fossil bone over the next few years. They came to the conclusion that it was, in fact, the rib bone from a Pacific white-sided dolphin. One scientist jokingly nicknamed the creature "flipperpithicus," meaning the dolphin ape.

We need to take care that we don't become overly impressed by the conclusions of scientists with lots of letters behind their names. When it comes to supporting evolution, many scientists are driven more by their desire to prove evolution than by any facts.

Prayer: Father in heaven, strengthen my faith so that I know what I believe and why I believe, so that I may not be intimidated by the claims unbelievers make against the truth of Your Word. In Jesus' Name. Amen.

Ref: W. Herbert. "Hominids Bear Up, Become Porpoiseful." *Science News*, Vol. 123, p. 246.

Another Miracle of Sight

Psalm 94:8-9
"Understand, ye brutish among the people: and ye fools, when will ye be wise? He that planted the ear, shall he not hear? he that formed the eye, shall he not see?"

The eye is a chemical and technological wonder. It would take thousands of carefully directed mutations for an eyeless creature to develop sight. And since a partial eye doesn't provide sight, each of these thousands of mutations would offer no advantage to the creature.

The fact is, of course, that mutations don't happen this way. All the true mutations we know about have harmed rather than helped the creature with the mutation. This fact places the thought of thousands of positive mutations in a row outside the realm of scientifically reasonable consideration. In addition, neither living nor known fossil animals show any evidence of gradual development of eyes. Therefore, the most reasonable scientific conclusion is that there is a Creator Who designed and created the eye.

The lens of the eye is a marvel of chemistry. It is made up of a concentration of protein molecules inside water-lined cells. When scientists learned this, they were amazed. Protein molecules in water are not transparent, as the lens must be. After more research, they discovered God's secret. The high concentration of protein molecules in the lens of the eye causes the proteins to pack together something like the molecules of window glass. As a result, the normally opaque protein solution in the lens becomes transparent.

The eye does more damage to ideas about origins that leave out God than almost any other feature of the creation because it allows us to see the Creator's fingerprints all around us!

Prayer: I thank You, dear Lord, for the gift of sight by faith that allows me to see the clear evidences of Your handiwork all around me. Teach me to show others, whose sight may not be so clear, that You are not only our Creator, but our Savior. Amen.

Ref: I. Peterson. 1983. "Why the Eye Lens is Transparent." *Science News*, April 16.

An Amazing Australian Frog

Mark 10:32a
"And they were in the way going up to Jerusalem; and Jesus went before them: and they were amazed; and as they followed, they were afraid..."

A rare Australian aquatic frog has one of the most amazing systems for rearing its young to be found in the animal kingdom. If any other creature tried the same method, the result would be fatal to its young.

The frog is a species of Australian frog so rare that it only has a Latin name. Once the eggs are laid and fertilized, the female swallows the eggs. Sometimes she waits until they begin to develop. Once the young are in her stomach, she forgoes eating for eight weeks. After eight weeks, the young, developed frogs emerge from her mouth.

Why aren't the young frogs digested in the mother's stomach? Once the offspring are in her stomach, they begin to release fine threads of chemicals from their mouths. Among these chemicals is one that causes the mother's stomach to stop manufacturing acid. In other words, the tiny young frogs change the mother's stomach from an organ for digestion into a comfortable, protected nursery. Scientists were amazed at the frog's system, since no other creature in nature is able to do this.

The Bible often notes that when Jesus Christ walked this Earth in bodily form, He frequently amazed people. Sometimes He used a miracle to astonish them. Sometimes it was His teaching that astounded them. Here we see that through His work as Creator, He can still amaze even the most hardened unbelievers with His work. You can learn more about why He is so amazing in the pages of the Bible.

Prayer: Dear Lord, I recommit myself to regular reading and study of Your Word. Through Your Word, strengthen my faith and help me to be filled with amazement over Your great love for me through the forgiveness of my sins. Amen.

Ref: "Tadpole Role in Gastric Pregnancy." *Science News*, Vol. 123, p. 350.

Destruction from Space?

Luke 21:27-28
"And then shall they see the Son of Man coming in a cloud with power and great glory. And when these things begin to come to pass, then look up, and lift up your heads; for your redemption draweth nigh."

We hear many claims that all life on Earth could someday be wiped out in a catastrophic collision with a giant asteroid.

According to Arthur Humphrey, immediate past president of the American Institute of Chemical Engineers, the Earth has just missed several catastrophic collisions with asteroids in the last few years. He said that in 1989 and 1991, two asteroids missed the Earth by only a few hundred thousand miles. Humphrey estimates that there are about 11,000 large asteroids that cross our orbit.

He suggests that scientists map all the asteroids that cross Earth orbit, giving us a chance to divert any that might threaten the Earth. The estimated cost of the project is $20 million.

The world is a terrifying place to those who think that our future lies in the hands of chance. But consider the design of our solar system. The Earth is just at the right distance from the sun for us. If we were one percent closer or further, we could not live on Earth. Our sun is a remarkably stable star. If it were like many stars, its energy could fry us one day and leave us to freeze the next.

These remarkable relationships are not a product of chance. They are the evidence of our Creator's loving design. And He has told us that human beings will be on Earth to greet Him when He returns.

Prayer: Dear Heavenly Father, I trust Your constant and loving provision and protection. Help me to see that when I am filled with the fear of some danger in the creation, I have been seduced by the pagan belief in chance that surrounds me. Forgive me for my fear for Jesus' sake. Amen.

Ref: Arthur E. Humphrey. 1991. "The Squeaky Wheel." *AICHE*. Dec., p. 12. K. M. Reese. 1992. "The Hazard of Asteroids." *C & EN*, Feb. 3, p. 68.

Can Science Beat Death?

Hebrews 9:27
"And as it is appointed unto men once to die, but after this the judgment..."

Are we programmed to die? Scientists who have been studying the aging process hold one of two positions.

One school of thought says that death takes place when too many of our body's parts simply wear out from use. The other school says that death is programmed into our genes. As a result, they believe that there are built-in limits to how long we can live.

What some scientists have called an "anti-aging gene" makes a chemical, abbreviated as SOD. SOD keeps the body from burning itself up. SOD is among a number of chemicals in the body that take the dangerous products of normal chemical reactions in the body and make them harmless. If not rendered harmless, these chemicals, called free radicals, can damage parts of our cells, including DNA. Through selective breeding, biologists have developed fruit flies that live twice as long as normal. Their studies show that these flies have a gene that makes higher amounts of SOD, and so seem to have extra protection against aging.

There is only one permanent solution to death. Death is the result of sin. Jesus Christ, Who has taken away the reign of sin, promises eternal life to all who come to Him for the forgiveness of sins. He lived the sinless life that we cannot live. Then He went before God with our sins, to take our punishment for those sins. Are you sure of the forgiveness of your sins through Jesus Christ? You can be!

Prayer: I thank You, dear Lord, that You gave up heaven and Your perfect union with the Father to make the forgiveness of my sins possible. I trust Your Word that You have earned my forgiveness. Help me now glorify You with my life. Amen.

Ref: Ronald Kotulak & Peter Gorner. 1992. "Fruit Flies Provide Hint on Aging." *Chicago Tribune*, Feb. 9, pp. 21, 26. San Francisco Examiner. 1991. "Does a Gene Program us to Die?" *Star Tribune*, Nov. 30.

Lightning-Like Vision

Matthew 13:15
"For this people's heart has waxed gross, and their ears are dull of hearing, and their eyes they have closed; lest at any time they should see with their eyes, and hear with their ears, and should understand with their heart, and should be converted, and I should heal them."

Your eyes are quicker than anyone ever thought.

For you to see an image, a huge number of chemical and electrical reactions must take place in sequence. Science still does not fully understand all the reactions. Science does know that each set of chemical or electrical reactions must take place in a sequence that leads to the next set of reactions. Obviously, each of these reactions in the chain must take place extremely rapidly for us to see what is happening while it is still happening.

Researchers have been studying how quickly light causes the first chemical changes in the eye that finally lead to you seeing an image. This type of chemical change is called a photochemical reaction. Photochemical reactions are the basis of photographic prints. However, the photochemical reactions that result in a printed photograph take place much more slowly than the photochemical reactions in your eye. Now, for the first time, scientists have timed the first photochemical reaction in the eye. They have found that the eye's photochemistry is among the fastest ever studied. They report that the first reaction takes place in 200 thousandths of one thousandth of a millionth of a second!

Clearly, the many chemical and electrical operations involved in sight could not have developed by trial and error, step by step, over huge spans of time. Our Creator has given us the ability to see so that we could see His handiwork in the creation. Even the process by which we see clearly shows the excellence of His work!

Prayer: Lord, thank You for so excellently designing our ability to see. Make me better able to help others see not only Your handiwork in the creation, but also the fulfillment of our need for the forgiveness of sins in You. Amen.

Ref: "Vision Event Occurs with Blinding Speed." *C & EN*, Oct. 21, 1991, p. 20.

Stars Are Dying Too Fast

Job 9:7-10
"Which commandeth the sun, and it riseth not; and sealeth up the stars. Which alone spreadeth out the heavens, and treadeth upon the waves of the sea. Which maketh Arcturus, Orion, and Pleiades, and the chambers of the south. Which doeth great things past finding out; yea, and wonders without number."

The Bible tells us that the entire universe is decaying and dying because of the burden of sin. We know from Scripture that the entire heavens and Earth as we know it will someday be destroyed by fire to make way for a new creation. The Bible also indicates that the entire creation is young.

These facts were generally accepted until evolution became a fashionable theory. Early evolutionists were quick to realize that people would not believe that a relatively young Earth could produce the life we see today. Chance and a young Earth simply don't go together. As a result, the Earth and the universe have been declared to be billions of years old.

Now new conclusions about the rate at which stars are dying can be added to the evidences that contradict an ancient age for the creation. When a star dies, it typically explodes, leaving a shock wave of expanding dust and gas. While stars are dying, no one has ever confirmed the formation of new stars. The fact that stars are evidently dying faster than they are being formed has been a problem for evolutionists. That problem recently became worse with the announcement that stars are evidently dying at twice the rate astronomers thought. Astronomers report that they have now found evidence of 24 stars that have died in the last several thousand years in a space of about only one-eighth of our galaxy.

What the Bible says about the birth and death of the creation is true. The evidences for both are all around us.

Prayer: Dear Father in heaven, as I face the truth of that day when the creation shall end, comfort me with Your sure promise of my salvation through the forgiveness of my sins in Your Son, Jesus Christ. In His Name. Amen.

Ref: "Stars Blinking out Faster than Once Believed, Scientists say." *The Vancouver Sun*, Jan. 11, 1992, p. 7.

Self-Esteem and Forgiveness

Luke 18:13
"And the publican, standing afar off, would not lift up so much as his eyes unto heaven, but smote upon his breast, saying, God be merciful to me a sinner."

We hear a lot about "self-esteem" today. The "self-esteem movement," made up of some religious teachers and psychologists, seeks to make people feel better about themselves without making any reference to sin or the need for forgiveness. Some religious teachers have even said that Christianity should stop talking about sin. Because the movement is associated with psychology, many mistakenly believe that the claims of the "self-esteem movement" have a scientific basis. They do not.

Consider the publican who prayed, "God, I thank you that I am not like other men…" From the world's standpoint he had no trouble with self-esteem. On the other hand, the world would say that the sinner who prayed, "God be merciful to me, a sinner…" had a serious self-esteem problem. In Jesus' analysis, however, the publican's self-righteousness – his "good self-image" – was what kept him from God. It was the sinner's knowledge of his self-worthlessness, and humble repentance, that brought him the peace of God in a personal relationship with his Maker.

From this vantage point it is easy to see that the cult of self-esteem promotes self-righteousness. Adding Christ to self-esteem still produces self-righteousness.

Christians and all people need to be encouraged to focus on Christ! The Christian who lives a daily life of repentance in the full knowledge that Christ has redeemed him, making full atonement and peace with God, will have no "self-esteem problem."

Prayer: Forgive me, dear Lord, for those times when I have thanklessly felt sorry for myself or let my pride come between us. Fill me with the joy and peace that only You can provide through the forgiveness of my sins. Amen.

Ref: Scott M. Marincic. 1992. "Grace and Truth – Not Self-Esteem." *Lutheran Witness*, Jan., p. 21.

Pawpaw Surprise

Job 5:11
"To set up on high those that be low; that those which mourn may be exalted to safety."

The mechanism God designed to protect the fruit of the pawpaw tree from being eaten before its seeds are ripe may provide powerful, new anti-cancer drugs and an effective, safe, insecticide. The pawpaw tree, with its custard-like fruit, is common all over the eastern United States.

The twigs of the pawpaw contain powerful biological chemicals that make them toxic to insects. The pawpaw fruit is nearly as toxic as the twigs until it ripens. This discourages animals from eating the fruit until it is ripe and the seeds are ready to be spread. Scientists at Purdue University recently developed a simple and inexpensive test to see if natural substances are biologically active. If a plant extract is active, it means that it could have some use as a medicine.

As a result of these successful tests, scientists found at least seven chemicals that may prove useful against cancer. The most powerful is a million times more effective against cancer than the leading anti-cancer drug in use today. In addition, they discovered a powerful, natural, insecticide that is effective against a wide range of plant pests.

God's extraordinary care for the lowly pawpaw also appears to provide powerful help for some of man's ills. God often chooses to work through the lowly things of this world to bring blessings to us. Nothing or no one is unimportant to His plans if they will only allow Him to use them.

Prayer: I thank You, dear Father, that You have chosen to work through the lowly things of this world so that even the least among us is not without Your help or Your blessing. Show me how You can use me to tell others of Your help in Christ. In His Name. Amen.

Ref: "Look What's Hidden in the Pawpaw." *Science News*, Feb. 29, 1992, p. 143. "Pawpaw Tree Yields New Cancer Drug, Pesticide." *Sacramento Union*, Feb. 8, 1992.

Millions of Dollars Per Pound

Psalm 139:14
"I will praise thee; for I am fearfully and wonderfully made: marvellous are thy works; and that my soul knoweth right well."

Sometimes information can be so simplified that it becomes deceiving. All of us have heard that the human body is made up of only a few dollars worth of chemicals. However, that commonly repeated statement doesn't reflect what's really in the human body.

Take the complex sugar called hyaluronic acid. This sugar, found in the bodies of all animals and man, is the most effective lubricant known. It is also elastic, which means that it can absorb shock and return to its original shape. Hyaluronic acid lubricates your joints. And it protects delicate tissues in your eyes.

Hyaluronic acid is added to some cosmetics, skin moisturizers, and shaving creams. It's also used in eye surgery to help the eye keep its shape and to protect eye tissues. The problem is, the hyaluronle acid used in eye surgery costs millions of dollars per pound. (Luckily, only a couple of hundred dollars worth is needed.) This high cost prevents another medical use for this amazing substance. Medical researchers believe that lubricating arthritic joints with hyaluronic acid would restore lubrication and ease the causes of joint pain. Research continues on less expensive methods for making hyaluronic acid.

When people forget how fearfully and wonderfully God has created us, it's easy to reduce the wonders of the value of the human body to a few dollars worth of chemicals. Surely this is not the thankless attitude we should express to others.

Prayer: Lord, I thank You that I am indeed fearfully and wonderfully made. Forgive my careless thanklessness and help me to more clearly see the countless reasons I have to continually thank and praise You! Amen.

Ref: Andrew Pollock. 1992. "Tapping the Body's Own Lubricant at Millions of Dollars a Pound." *The Vancouver Sun*, Feb. 8, p. 7.

What Meteorologists Think About "Greenhouse"

Matthew 24:30
"And then shall appear the sign of the Son of Man in heaven: and then shall all the tribes of the earth mourn, and they shall see the Son of Man coming on the clouds of heaven with power and great glory."

What do professional meteorologists think about claims that the Earth is undergoing a greenhouse effect because of man's activities? Scientific opinion is growing that there is nothing to the idea that the Earth is either warming or in danger of warming under the proposed influence of the greenhouse effect. More in the scientific community are also becoming increasingly outspoken in their opposition to the false science used to support the idea that man's activities are causing the Earth to become warmer.

The Gallup poll recently interviewed a cross-section of members of the American Geophysical Union and the American Meteorological Society. Do these scientists, who should know better than anyone else, think that the greenhouse effect is something to fear? According to the poll, *less than half* – 41 percent – think that there is scientific evidence for global warming. However, 70 percent thought that the scientific work supporting greenhouse conclusions is "fair to poor."

So it is not true that all scientists accept the idea that greenhouse warming is going on. Less than half of the professional meteorologists believe it's happening. And most suspect the quality of the scientific work supporting ideas about greenhouse warming.

While Christians favor good stewardship of our resources, we need not fear the destruction of our planet, as do the unbelievers. Christ clearly tells us that Christians will be here to greet Him when He returns.

Prayer: I thank You, Father, that You have promised to protect and provide for Your people on Earth until Christ returns. I ask Your forgiveness for those times when I have carelessly misused and abused Your creation. Amen.

Ref: "Greenhouse Deconstructed." *National Review*, March 16, 1992, p. 17. Peter Samuel. 1992. "Hard Science Ices Global Warming Hysteria." *The Detroit News,* Feb. 17, p. 11.

Quick Coal

2 Peter 3:3-4
"Knowing this first: that there shall come in the last days scoffers, walking after their own lusts, And saying, Where is the promise of his coming? for since the fathers fell asleep, all things continue as they were from the beginning of the creation."

How is coal made? How long does it take? Scientists did not know the answers to these questions until the last few years. Despite this lack of knowledge, textbooks have taught for generations that it takes millions of years to make coal. This was another supposed proof that the Bible's view of history is wrong.

In recent years, scientists who believe in creation showed that coal could be formed in much less time. They also showed how the coal beds offered evidence that they were formed rapidly. Unfortunately for science, much of the scientific community ignored them because they were challenging claims made by evolution.

In the last ten years, evolutionists in the scientific community have begun to produce their own work that confirms what creationists have been saying all along. For example, scientists at the Argonne National Laboratory heated lignin, the "glue" that holds the fibers in wood together, to 300 degrees Fahrenheit in the presence of clay. This temperature is fairly common in geological formations, and coal is found with clay. Even though the lignin was heated only for between two weeks and a year, coal formed. The lignin heated for a year produced high-grade coal. No, millions of years were not necessary.

This research further strengthens creationist research claims that the flood at the time of Noah, only about 4,500 years ago, is most likely responsible for most of the world's coal beds. Again, the false scientific claims that challenged the truth of Scripture have been discredited.

> *Prayer: I thank You, Lord, that even man's best and most educated ideas are shown to be lies when they challenge the truth of Your Word. Help me not to become a scoffer against Your Word by failing to study or apply it in my life. Amen.*

Ref: "Ranking Theory Over the Coals." *Science News*, Aug. 6, 1983, p. 93.

Natural Human Language

Genesis 1:28
"And God blessed them, and God said unto them, Be fruitful and multiply, and replenish the earth, and subdue it: and have dominion over the fish of the sea, and over the fowl of the air, and over every living thing that moveth upon the earth."

According to evolution, human language began to develop when ape-like creatures started to grunt and make other noises at each other. Eventually they agreed on meanings for these sounds and language was born. According to the Bible, language is a gift of our Creator, Who is the source of all language. He made us to have fellowship with Him. He gave us language as an important part of our relationship with God.

The evolutionary view of language doesn't explain why the more ancient languages are the most complex, but the biblical view does. The biblical view also explains why research is showing that we are programmed for language.

In one instance, researchers studied the communication of profoundly deaf children with hearing parents, neither of whom had ever been taught sign language. These children had never heard or seen normal language, so they wouldn't be influenced by their surroundings. Researchers wanted to find out if the children followed a pattern in how they put their communication together. To their surprise, they found that the children all followed the same patterns of sentence structure. The pattern used by the children is one of two patterns into which all the world's languages can be divided.

This research supports the Bible's claim that we were created with a built-in ability for language.

Prayer: Dear Father, I confess that I have abused Your gift of language with foolish and harmful speech and by not using language to pray to You as often as I could. For Jesus' sake, forgive me and help me to use language as You intended. Amen.

Ref: P. D. Sackett. 1983. "Do Deaf Children Show Language Learning Bias?" *Science News*, July 30, p. 73.

Stars That Are Too Fast?

Revelation 21:23
"And the city had no need of the sun, neither of the moon, to shine in it: for the glory of God did lighten it, and the Lamb is the light thereof."

The stars are among the most beautiful and powerful evidences for our Creator. While there are billions of them, no two seem to be identical. Even so, they can be classified into groups.

The hottest stars are blue. Our own star, the sun, is an average star. It is classified as a yellow star. Red giants can be more than a hundred million miles in diameter. One of the best known red giants in the night sky is Betelgeuse, which is 18,000 times brighter than our sun. Good thing it's 300 light-years away!

Scientists say that in general, the hot blue stars are young. They say that yellow stars, like our sun, are middle-aged and when they get old, they will turn into red giants before exploding. They say these changes take billions of years. But their evolutionary scheme for determining the age of stars recently suffered a serious blow. Astronomers report that a star called *FG Sagittae* went from a blue star to a yellow star in only 36 years! This calls in question their entire evolutionary scheme for dating stars with vast ages.

According to the Bible, the entire universe is only about 6,000 years old. And since man brought sin into the world, everything, including the stars, is running down. But we need not worry about our sun swelling up, exploding, and killing all life on Earth. The Bible tells us that it will remain until our Lord returns for us. And in the new creation, there will be no need for the sun.

Prayer: *I thank You, Lord, for the warmth and beauty of the sun and the life it supports on our Earth. I eagerly anticipate the new heavens and Earth when You shall be the only light that I need. Amen.*

Ref: "Speedy Star Sequence." *Creation Ex Nihilo*, Vol. 14. No. 1, p. 7.

Stone-Making Plants

Psalm 30:12
"To the end that my glory may sing praise to thee, and not be silent. O LORD my God, I will give thanks to thee for ever."

Have you ever wondered how you could get cut by a blade of grass or the edge of a palm leaf? Plants have a nasty secret. They absorb silica, the same stuff of which glass is made, and store it in their cells.

Some plants, like corn, store large mounts of silica on their leaves, making the edges of the leaves sharp and strong enough to give you a nasty slash. And yes, they do this largely to protect themselves. Insects that like to eat plant leaves find that when they chew on plants with lots of silica, their mouth parts wear out faster. Plants also store silica between their cells, giving them a strong, stony skeleton.

These silica particles made by plants are called *phytoliths,* which literally means "plant stones." Chemically, they are colorless and transparent opals. They can range in size from a thousandth of a millimeter to a millimeter. Each plant forms *phytoliths* that are unique in shape, and some plants make up to a dozen different types. After the plant has died and decayed, the *phytoliths* and their unique shapes remain in the soil. This means that with a microscope and a good knowledge of which plants make which *phytoliths*, you could examine the soil and tell which plants have grown there in the past. This is how scientists learned that the Indians were growing corn in South America 2,500 years before the birth of Christ!

God's creation is filled with wonderful surprises. And every one of them glorifies Him!

Prayer: I give You thanks, Lord, for Your wonderful creation. Forgive me for those times I am slow to thank You or when I take Your wonderful works for granted. Help me see more of the wonders You have made. Amen.

Ref: Ivars Peterson. 1983. "Plant Stones." *Science News*, Vol. 124. Aug. 6, pp. 88-94.

The First Are Made Last

Matthew 6:25
"Therefore I say unto you, Take no thought for your life, what ye shall eat, or what ye shall drink; nor yet for your body, what ye shall put on. Is not the life more than meat, and the body than raiment?"

The hard-driving, powerful executive who's in charge at all times is one of the most common images of success in the modern world. There is no end to the books and videos that are designed to help people rise above those around them by gaining power.

Power-motivated people tend to get into stressful situations. Research shows that as they create stress for those around them in their drive for power, they may be doing *more* harm to themselves. Research has repeatedly shown that stress lowers the body's immunity to disease. Other studies have linked stress to serious health problems and early death.

In one study, researchers classified test subjects into those who are motivated by close relationships with other people and those who are motivated by power. Both groups showed lower immunity during times of stress, as expected. The power-motivated subjects, however, consistently showed a much greater decrease in their immunity as a result of stress.

The world values those hard-driving people who gain power over others. But research on stress consistently shows one way in which these people are in truth not very successful in the true business of life. It appears that a life of selfless service to others in the Name of Christ not only benefits the Kingdom of heaven, it also benefits humble servants in their lives right here on Earth!

Prayer: Lord, take my life and let it be one of humble service to You. Let me follow Your example in giving myself, rather than the world's values of seeking power and control over people, so that my life may truly glorify You. Amen.

Ref: P. Taulbee. 1983. "Study Shows Stress Decreases Immunity." *Science News*, July 2, p. 7.

Monkey Medicine

Matthew 6:28-29
"And why take ye thought for raiment? Consider the lilies of the field, how they grow; they toil not, neither do they spin: And yet I say unto you, that even Solomon in all his glory was not arrayed like one of these."

More than a quarter of our drugs come from wild plants. Traditionally, researchers in search of new drugs would ask local folk healers about plants that might be helpful. That isn't always the best approach. In some societies, plants are thought to have healing powers simply because they look like a disease.

Researchers are now learning to follow animals around in search of natural medicines. Monkeys have led medical researchers to several potentially new medicines. Chimps in Tanzania were seen swallowing the unchewed leaves of a tree. Studies of the leaves and the droppings showed that the leaves were not digested. The chimps' digestive system simply removes chemicals from the surface of the leaves that kill intestinal parasites. The female of a rare species of woolly spider monkey in Brazil does not ovulate for about six months after she gives birth. This gives her time to raise her offspring. About six months after giving birth, the female will travel outside its usual area to gorge on the fruit of a specific tree. Researchers found that its fruit is rich in a chemical that causes the female to produce a hormone that re-starts ovulation.

As in other cases of this type, we have to ask the question: Do these animals really have innate medical knowledge, or is it that God cares for them when they are sick and leads them to the right medicinal plant?

Prayer: Dear heavenly Father, I thank You that there is no limit to Your goodness and generosity to all Your creatures. Help me to remember Your love and Your generosity when I am tempted to worry about the future. In Jesus' Name. Amen.

Ref: Boyce Rensberger. "Researchers: Chimps Use Herbal Medicine."

Mistletoe Mimicry

1 Peter 5:6-7
"Humble yourselves therefore under the mighty hand of God, that he may exalt you in due time: Casting all your care upon him, for he careth for you."

We usually identify plants by their appearance. Palm trees can be identified because the many different kinds of palms have some common identifying characteristics. Pine trees, likewise, have common features that help us identify them. One plant has become a master at looking like many other plants. It has a good reason for doing this.

Mistletoe is a parasite that's commonly found in Europe, North America, and Australia. While it uses its green leaves to make its own food, it gets its water and minerals through roots attached to a host. It's relatively easy to see the difference between American and European mistletoe and its host. However, many species of Australian mistletoe mimic the host on which they grow.

The drooping mistletoe is so named because its leaves look like those of its host, the eucalyptus tree. The box mistletoe and the pendulous mistletoe both have hard, sickle-shaped leaves that make them look much like other eucalyptus trees on which they grow. Botanists who believe in evolution are divided on how to explain this mimicry. Mistletoe can neither see its host, nor change form like an amoeba.

There's no problem, however, if we understand that our unlimited Creator also cares about the living things He made. Mistletoe that looks almost identical to its host can make its living without attracting attention to itself.

Prayer: Father, I am filled with wonder, and I thank You that Your creativity and love are so evident in the creation. You generously provide for all Your creatures. I especially thank You that You have provided for my greatest need, the forgiveness of sins, in the life and death of Your Son. Amen.

Ref: Peter Bernhardt. 1989. *Wily violets and underground orchids: revelations of a botanist.* William Marrow and Company, Inc., pp. 34-37.

The Parting of the Red Sea

Exodus 14:21
"And Moses stretched out his hand over the sea; and the LORD caused the sea to go back by a strong east wind all that night, and made the sea dry land, and the waters were divided."

The parting of the Red Sea as the Israelites escaped Egypt is one of the most attacked miracles in the Bible. Much of scholarly opinion says that the Israelites didn't really escape through the sea. They simply went through a nearby marsh, called the Reed Sea. In the first place, how could the Israelites, loaded down with their own possessions plus the spoils of Egypt, have traveled through a marsh? Second, how could an entire army drown in a swamp?

Now an article in the bulletin of the American Meteorological Society shows that the Red Sea could have parted just as the Bible describes. In making their case, meteorologists used sophisticated computer simulations and the Bible's description of the event. Because the extension of the Gulf where the Israelites crossed is so long and shallow, the strong east wind described in Exodus 14 could have lowered the water level by ten feet. The Israelites could then cross on an underwater ridge, with water on both sides of them. The water backed up into the wider portion of the Red Sea could return within a matter of minutes after the wind stopped, drowning pharaoh and his army.

So many of the miraculous events described in the Bible have been explained away by science as natural events, and this may just be another case. But all this means is that God can use natural events and that He is in control of when and where they occur.

Prayer: Dear Father in heaven, I thank You for those dedicated scientists who take Your Word seriously and use their talents to glorify You. Let their witness be heard, and move more of our Christian young people to study for the sciences. In Jesus' Name. Amen.

Ref: Thomas H. Maugh II. 1992. "A Miracle Adds Up." *Star Tribune*, Mar. 14, p. 16Ae.

Unscientific Source?

Psalm 119:140
"Thy word is very pure: therefore thy servant loveth it."

Most Christians have been told, "Creation isn't scientific because it comes from the Bible!" Does the source of an idea make the idea unworthy of science?

Many believe that the only truly scientific ideas come to people dressed in white smocks working in laboratories. Experience shows, however, that scientific discoveries can come in the most unusual ways. Friedrich Kekule discovered the theory of atomic valency in a vision while traveling on the top deck of a London bus.

A new, natural insecticide was also discovered in what might be considered an unscientific fashion. Mechanics working at a University of Georgia field research station used a soap called Dirt Squad. Dirt Squad is a grease cutter made from orange peels. The mechanics had dumped some of the cleaner on a fire ant hill. Listeners who are familiar with fire ants know that these nasty members of the insect kingdom are hard to kill. This cleaner, though, killed all the ants. Hearing this, one of the scientists used the cleaner on his cat, who had fleas. The soap killed all the fleas. This led scientists at the research station to investigate the oils in citrus peels. They discovered a natural chemical in citric oil that is harmless to vertebrates but deadly to insects.

Any idea, from any source, including the Bible, needs to be considered before it can be accepted or rejected as scientific.

> **Prayer: I thank You, Lord, that Your Word is worthy and true. I also thank You that the Bible tells us about much more than can be tested by our experience and science. And I thank You for those who are showing that where the Bible talks about things that can be studied by science, Your Word stands true! Amen.**

Ref: J. A. Miller. 1983. "An Insecticide with a Twist." *Science News*, Oct. 8, p. 231.

Purpose? To Glorify God!

Psalm 115:1
"Not unto us, O LORD, not unto us, but unto thy name give glory, for thy mercy, and for thy truth's sake."

God has created us with a strong desire to know the purpose of things. Evolutionists generally believe that if a feature doesn't serve a purpose, it should not have evolved. On the other hand, the Bible tells us that purpose is given by God. He gives the woodpecker its incredible tongue because it helps the woodpecker to make its living. In addition, this wondrous design glorifies its Creator. At other times it seems that the Creator has produced some living thing for which there is no apparent purpose other than to glorify Himself. These are the features that give evolutionists the most difficulty.

Male antelopes show a vast array in the designs of their horns. Males may have spears, spirals, curves, twists, or simply short daggers. These differences in design suggest that differences in purpose caused the horns to evolve, but that doesn't make sense. Even stranger is the fact that female antelopes generally have smaller horns that are much more like daggers. Evolutionists are stumped about why female antelopes would evolve short horns while males evolved such a variety.

As Bible-believing Christians, we have no difficulty accepting that the variety of designs in male horns, and the similarity in design of female horns, all serve to glorify the Creator and show that evolutionary explanations don't really explain anything.

> *Prayer: I thank You, dear Father, for the variety and beauty You have created in the living world. I also thank You that these features glorify You in so many ways. Help my life and words glorify You and bear clear witness to the salvation You have provided for us in the forgiveness of sins through Your Son, Jesus Christ. Amen.*

Ref: J. A. Miller. 1983. "Antelope Horns: Female Perspective?" *Science News*, Sept. 17, p. 183.

Intelligent Animal Antics

Psalm 71:17
"O God, thou hast taught me from my youth: and hitherto have I declared thy wondrous works."

While debate continues about whether chimps can actually learn to communicate with language, there is little question that they understand simple sentences. However, this doesn't prove any relationship between man and apes. Research shows that dolphins and sea lions can also learn language. Researchers used the same methods to get people and pigeons to memorize a string of numbers. They found that people can usually remember no more than nine numbers in a sequence; pigeons can remember five. They also discovered that pigeons, like people, memorize lists by grouping together similar elements in the list.

Scientists always thought that the plover's "wounded wing" display was an instinctual reflex to lead predators away from its young. New research shows that the birds behave differently depending on whether they are familiar with an intruder and know if the intruder is a true threat. It's been discovered that dolphins try to hide from tuna boats by not jumping when a tuna boat is near. If this doesn't work, dolphins will swim on the right side of the boat because the cranes and nets are usually on the left side.

As one scientist put it, animals are as smart as they need to be, no more and no less. When God made the animals, He was generous with something He has an unlimited amount of – intelligence.

Prayer: I thank You Lord, that You are generous with all Your gifts. Forgive me for those times when I have misused or belittled the intelligence You have given me. Help me to better use my intelligence to spread Your kingdom. Amen.

Ref: Shannon Browniee. 1985. "A Riddle Wrapped in a Mystery." *Discover*, Oct., pp. 85-93.

Dissolving the Magic of Hypnosis

Philippians 2:5-7
"Let this mind be in you, which was also in Christ Jesus: Who, being in the form of God, thought it not robbery to be equal with God: But made himself of no reputation, and took upon him the form of a servant, and was made in the likeness of men:"

At Creation Moments we are often asked about the safety of hypnosis. Some law enforcement agencies regularly make use of hypnosis. It has been widely believed that hypnosis helps people remember details about an incident that they don't naturally remember. Some judges even accept memories said to be enhanced by hypnosis as evidence in court. However, the claims made for the powers of hypnosis have seldom been tested scientifically. This led researchers from the University of Waterloo in Ontario to test whether hypnosis enhances memories or actually creates memories.

The researchers showed test subjects sixty drawings of common objects and then asked them to remember as many as possible. Those who were hypnotized remembered only slightly more of the pictures. At the same time, they made three times as many mistakes as the unhypnotized subjects in "remembering" pictures they never saw. Scientists also found that people most susceptible to hypnosis were most likely to report false memories. They concluded that hypnosis appears to trick the mind into recognizing unreal "realities."

Christians are wise to avoid hypnosis. Its dubious claims of benefits are not supported by science. Worse, hypnosis appears to leave the mind open to influences that can be harmful.

> **Prayer: Dear heavenly Father, I know that there are enough negative influences in the world without exposing myself to more dangers. I pray that You would help me to be more discerning of the influences around me and better able to wisely evaluate them in light of Your Word. In Jesus' Name. Amen.**

Ref: "Memories are made of this." *Science News*, Vol. 124, p. 248.

Fly Eats Toad!

Psalm 104:27
"These wait all upon thee; that thou mayest give them their meat in due season."

It's not news when a dog bites a man. It is news when a man bites a dog. Likewise, it's not news when a toad eats a fly. But news about a fly that eats toads should get our attention.

The unusual drama of a toad-eating fly was first discovered in the Arizona desert by Thomas Eisner of Cornell University. He noticed a muddy desert pond that had a large population of spadfoot toads. Looking closer, he noticed that many of the toads were in distress, and some were being pulled down into the mud at the bottom of the pond. A little digging revealed that the predator was the larvae of horseflies.

Further study showed that the larvae burrow into the soft mud until only their head is barely exposed. When an unsuspecting toad wanders by, that larva grabs the toad with its powerful mandibles and pulls it into the mud. The larvae inject their victims with venom and then consume the toad's body fluids. Eisner says that this drama plays out wherever there are horsefly larvae.

While this is not the prettiest subject we've ever covered, it certainly is one of the strangest. However, consider the Creator's design here. The horsefly larvae eat toads in order to grow into adults. The toads eat the adult horseflies to make more toads. It appears that God has designed a rather circular food chain that operates quite nicely in the desert where food is scarce.

> **Prayer: Dear Lord, I thank You that You have so generously provided for all of the needs of all of Your creatures, no matter what the circumstances. Help me to remember the individual care You give to all of Your creatures next time I am tempted to worry about the future. Amen.**

Ref: J. Greenberg. 1983. "For Whom the Bell Toads – Poetic Justice in the Arizona Desert." *Science News*, Nov. 5, p. 293.

Rise and Shine!

Psalm 57:8-9
"Awake up, my glory; awake, psaltery and harp: I myself will awake early. I will praise thee, 0 Lord, among the people: I will sing unto thee among the nations."

On the first morning of the time change to daylight savings, it's a little harder to get up. That night it might be a little more difficult to feel sleepy at bedtime. This is because our bodies need a little time to adjust to the time change.

All of us have a built-in, 24-hour biological clock that regulates our daily cycles of body temperature, hormones, and sleepiness and wakefulness. As the days get longer or shorter, or when we travel from one time zone to another, our clocks need to be reset. Medical researchers have found that insomnia is sometimes caused by a mis-set biological clock. Research shows that our clocks are reset by light. Your biological clock is a group of nerve cells in the hypothalamus at the base of your brain. How do those cells know whether they are keeping the proper time?

To find out, scientists kept rats in complete darkness for seven days. Tests with both humans and animals show that this will effectively confuse the biological clock. Then researchers exposed the rats to light just before what normally would be dawn. They report that the light caused immediate activity in two genes within the biological clock. This was the only part of the brain in which the genes responded to light.

Our biological clock with its reset mechanism is a practical invention of our all-wise Creator that inspires wonder even among those who don't believe in Him.

__Prayer: I praise You, dear Father, for all the wonders You have designed into my body. I also thank You that You have so wonderfully designed them that even those who do not believe that You are the Creator are filled with wonder over Your marvelous work. In Jesus' Name. Amen.__

Ref: M. Stroh. "Genes May Help Reset Circadian Clock." *Science News*, Vol. 141, p. 196.

Cubic Ice

Psalm 9:1
"I will praise thee, O LORD, with my whole heart; I shew forth all thy marvellous works."

We've all seen a misty halo around the sun or the moon. The halo is caused when light from the sun or the moon shines through ice, high in the atmosphere. Ice crystals, of course, are hexagonal, or six-sided. And for this reason, the halo always appears at an angle of 22 degrees away from the sun or moon.

However, all isn't as simple as it seems. A rare halo around the sun or the moon sometimes appears at 28 degrees instead of 22 degrees. But don't stay up late gazing at the full moon to see it. This halo, called Scheiner's halo, has been reported only seven times in the last 350 years. Scientists know that such a halo could be caused by octahedral, or eight-sided crystals. What crystals could cause this 28 degree halo? The mystery was recently solved by scientists working with cubic ice. Cubic ice is an unusual form of water ice. When water is frozen at temperatures of 100 below zero, it forms octahedral instead of hexagonal crystals. They used a little math to show that octahedral water crystals could cause Scheiner's halo. Scientists are now trying to discover whether conditions in the upper atmosphere could actually form cubic ice.

Every detail of the creation shows a tiny facet of the wondrous mind of God. Cubic ice shows us there is no limit to His imagination and creativity.

Prayer: I thank You, Lord, for the wonder that fills me as I learn about Your marvelous works. When I fail to understand why certain events happen in my life, help me to remember that Your imagination as well as Your power are unlimited. Help me always to place my trust in You rather than in what You have created. Amen.

Ref: "An Icy, Cubic Ring Around the Sun." *Science News*, Vol. 125, p. 8.

Magic Mirror

1 Corinthians 13:12
"For now we see through a glass, darkly; but then face to face: now I know in part; but then shall I know even as also I am known."

It's a widely accepted modern myth in our computer-driven, technological age that people today are in some way more advanced than people of thousands of years ago.

The biblical picture is that man has always been man. He has the same emotions, cares, and needs today that he has always had. He has always been as curious about his world as he is today. The biblical measure of man is not technological, but moral.

Most of these points are difficult to prove. Occasionally, modern man discovers ancient evidence that reveals amazing technology and inventiveness. The makyo, or magic mirror, was invented in China centuries ago. It's a mirror made of polished bronze that has a design cast into the back. When the mirror is used to reflect light onto a screen, the image on the back of the mirror appears on the screen. Why does the makyo do that? The first scientific paper on the subject was published in 1877. Modern scientists have studied the mirrors using the most sophisticated metal analysis methods known. Yet, they have not been able to figure out the mirror's secret. That nameless Chinese inventor who designed the makyo centuries ago was a genius, indeed.

Man is ever curious and inventive. The true measure for each of us centers on the forgiveness of sins that has been won for us by Jesus Christ on the cross of Calvary.

Prayer: Dear Father in heaven, forgive me for those times when I have forgotten that we are measured in terms of our sin and the forgiveness of sins through Jesus Christ, not in terms of our cleverness. In Jesus' Name. Amen.

Ref: Dietrick E. Thompson. 1984. "Inscrutable." *Science News*, Vol. 125, Jan. 14, pp. 30-31. See also http://videos.howstuffworks.com/science-channel/30250-what-the-ancients-knew-the-magic-mirror-video.htm.

The Unique Bdellas

Matthew 10:29
"Are not two sparrows sold for a farthing? and one of them shall not fall on the ground without your Father."

Scientists describe the creatures as mean, greedy, and antisocial. They are also impressive chemists. That's some reputation for a creature whose life span is only four hours. The short form of this microscopic creature's name is simply *bdella*.

Bdellas are predatory bacteria that live in fresh and saltwater as well as in sewage. In the first stage of its life cycle, it swims freely, using large whip-like flagella. The favorite food of the bdella is *E. coli* bacteria and, swimming ten times faster, the bdella rams the bacteria, punching a hole in its outer membrane and injecting six different enzymes. Then the bdella drops its flagellum and enters its prey, where it starts the second stage of its existence.

Over the next two or three hours, the bdella consumes the bacteria and reproduces. The new bdellas rupture the membrane and swim free to start the cycle all over. Scientists believe that the fact that bdellas eat so many different kinds of bacteria means that their purpose is to control the bacteria population.

It is interesting that even though many scientists reject the Creator, when confronted with an amazing creature like bdellas, they look for purpose. Truly, the whole creation glorifies our Creator.

Prayer: Father, I thank You that You are the Author of design and purpose. Forgive me for the times that I have talked about "luck" and "chance," and thus failed to give a good witness to Your love and personal care. In Jesus' Name. Amen.

Ref: Martha Wolfe. 1984. "Pee Wee Predator." *Science News*, Vol. 125, Jan. 28, pp. 60-61.

The Brooding Father Frog

Ephesians 6:4
"And, ye fathers, provoke not your children to wrath: but bring them up in the nurture and admonition of the Lord."

He sits on his eggs to protect them until they hatch, but he's no rooster. He's a good swimmer, but he's no duck or swan.

The Puerto Rican forest frog, coqui, is among the most common vertebrates in Puerto Rico. The frogs are active and hunt mainly at night because they are very susceptible to dehydration during the day. So important is protection from drying that researchers have found that the frog population is directly related to the number of sheltered spots they can find during the day. The frog's eggs, which are not laid in water, are even more threatened by dehydration than the adults. These frogs are important in the tropical forest because they are one of the forest's primary insect predators.

After she lays her eggs, the female leaves for parts unknown while the dedicated father tends the eggs. The average male coqui will sit on the eggs to keep them moist for 23 hours a day for up to three weeks. Without his care and protection, few eggs would ever hatch.

While this arrangement may seem strange at first, the genius of the design becomes apparent with a little thought. God designed these frogs to be major insect controllers in the forest. After she lays her eggs, the female's need for insects is much greater than the male's. It's easier for the male to give up most of his hunting for food for three weeks than it is for the female to give up eating. Again, our Creator's wisdom is evident!

> ***Prayer: Dear Father in heaven, I praise You for sharing a bit of Your loving and caring nature as our Heavenly Father with the father coqui. I pray that You would help Christian men do a better job of reflecting what it means to be loving, caring fathers with Your nature. In Jesus' Name. Amen.***

Ref: "Proud Papa Frog Protects Eggs." *Science News*, Vol. 125, p. 72.

Smart Bugs

Matthew 6:8
"Be not ye therefore like unto them: for your Father knoweth what things ye have need of, before ye ask him."

Can you teach an old insect, or even a young insect for that matter, any tricks at all? Science long assumed that insects were too stupid to learn even simple things. Researchers have now proven that insects can both learn and generalize knowledge into long-term lessons for life.

Researchers wanted to discover whether insects can learn, through a bad experience, to avoid certain bugs. They offered milkweed bugs to preying mantises. Milkweed bugs that feed on milkweed accumulate milkweed's poisons in their bodies. The mantises ate one bug, and then vomited it up. Each mantis refused a second helping of milkweed bug. They even rejected bugs that were painted to look like milkweed bugs.

A second test was run to test the response of preying mantises who had never had noxious milkweed bugs. The milkweed bugs they were offered were raised on sunflower seeds, so they would have no poison accumulation in their bodies. These preying mantises ate the bugs without getting sick. They continued eating the bugs when they were offered. These test mantises would only stop eating the bugs after they ate one that made them sick.

Learning from experience and later applying that knowledge to similar situations has always been considered a more advanced intellectual function. From the creation point of view, however, these abilities were given even to insects by the Creator because He knew they would need them.

> ***Prayer: I am comforted, Lord, by the fact You have left nothing in the creation without its needs supplied. When I am concerned about the future, help me to remember that if You have this much care for insects, You surely have even more care for me, since You gave Your life on the cross for my salvation. Amen.***

Ref: D. Franklin. 1984. "Spineless Predators 'Learn': Prey Can Cause Emesis in Nemesis." *Science News*, Mar. 17.

Seeking Rules

Deuteronomy 6:6-7
"And these words, which I command thee this day, shall be in thine heart: And thou shalt teach them diligently unto thy children, and shalt talk of them when thou sittest in thine house, and when thou walkest by the way, and when thou liest down, and when thou risest up."

The debate has raged for years. Some have said that as we grow up, we are seeking rules by which life is lived. Others have said that rules are bad and cause young people to rebel.

An eight-year national study of 2,000 adolescents should help settle the debate. Researchers studied the influence of family, community, and peer values among teenagers. Teens were selected from urban, suburban and rural settings.

They concluded that teens are "desperately" – their word – looking for the rules by which life is lived. They found that family and community values play a more important part of teens' lives than peer values. If parents are offering consistent values, teens will adopt those values before they will accept peer values. Researchers found that in urban areas there generally is a conflict between the values set by parents and those set by other elements in the community such as schools. Teens often responded to this disharmony by rejecting all rules and making up their own. Street gangs would be one of the most extreme examples of this reaction.

The Bible strongly encourages parents, and especially fathers, to teach their children. This research shows that young people are looking for that instruction from their parents. At the same time, young people need to see that the society their parents have created consistently follows the same values.

__Prayer: Lord, I thank You that my holiness is Your gift through the forgiveness of my sins, for my life fails to perfectly reflect Your values. As You have forgiven me, grant me Your Holy Spirit so that my life may more consistently reflect Your truth and power. Amen.__

Ref: "Growing Up: An Adult Connection." *Science News*, Feb. 18, 1984, p. 107.

Your Sixth Sense

1 Corinthians 8:1b
"We know that we all have knowledge. Knowledge puffeth up, but charity edifieth."

"Name the five senses." Every school child is asked that question. Now the question is outdated because a sixth sense has been discovered.

Years ago, University of Utah anatomist David L. Berliner noticed a strange reaction among coworkers. Normally his coworkers did not get along well. When he was working with open vials of extracts from human skin, however, there was a noticeable rise in friendliness. A few months later, when he sealed the extracts, coworkers were again unable to get along with each other.

That experience led Berliner, thirty years later, to discover the purpose of the vomeronasal organ – VNO for short. In animals, the VNO detects odorless pheromones – chemicals used to communicate readiness to mate and similar messages. Evolutionists thought that the human VNO was a useless vestigial organ left over from our evolution. Berliner discovered that the centimeter-long organ in our nose detects odorless human pheromones. It uses a system completely different from your nose. The VNO sends its messages to the hypothalamus, while the nose is wired to other sections of the brain. The hypothalamus controls basic drives and emotions.

Scientists long ago thought that they had identified all the senses. The discovery of the VNO shows that science still has very much to learn about the material creation. That places science in a poor position to pass judgments on biblical truth.

Prayer: Dear Father in heaven, I praise You that You have so wondrously made us. There is much to praise You for that we still don't know about. Help me not to be puffed up with false knowledge, but to humbly place my trust in Christ. Amen.

Ref: Gene Bylinsky. 1992. "A Sixth Sense That Affects How You Feel." *Fortune*, Jan. 27, p. 99.

Did Humans Evolve from Cockroaches?

2 Peter 1:16
"For have not followed cunningly devised fables, when we made known unto you the power and coming of our Lord Jesus Christ, but were eyewitnesses of his majesty."

Scientists who believe in evolution will argue that similarities between different species are the result of evolution. They say that similarities between humans and apes show that we are closely related. The fact that we are so unlike sponges or microbes, says this argument, shows that we have evolved much further.

The problem with this argument is that it often leads to absurd conclusions. Take, for example, the discovery that a chemical in the brains of cockroaches closely resembles two human hormones. Does that mean that humans evolved from cockroaches? Evolutionists are quick to deny an evolutionary link between humans and cockroaches. Still, they add that humans and cockroaches must have a common ancestor.

Or consider the protein abbreviated CK that is found in the brains of many species. Evolutionists would expect that the CK would be most similar in the most closely related species. From what evolutionists say, one would expect humans and apes to have the most similar CK. When scientists compared all the CK information they have, however, the two most closely related creatures turned out to be the domestic housefly and the African elephant!

When facts such as these are interpreted in terms of the theory of evolution, it simply flies in the face of reason, but this is the only option when the Creator God of the Bible is denied.

Prayer: Dear Lord, I ask that You would help me to identify those foolish claims made by men that call the truth of Your Word into question. Increase my thirst for the truth of Your Word. Amen.

Ref: Katie Tyndall. 1986. "Man and Roach. Pet Microchips." *Insight*, Nov. 3, p. 50; *Creation ExNihilo*, Vol. 14, No. 1, p. 37.

No Evolution in the Galapagos Islands

1 Corinthians 8:2
"And if any man think that he knoweth any thing, he knoweth nothing yet as he ought to know."

Ever since Charles Darwin first visited the Galapagos Islands, the creatures of these islands have been presented to the public as classic examples of evolution. Yet, evolutionists have long known that their claims could not possibly be true.

The little secret not included in the textbooks is that the Galapagos Islands are far too young for evolution to have produced the variety of life found there. They date the islands at only a couple of million inflated evolutionary years. They agree that this is far too short a time for evolution to have taken place.

For the last ten years, evolutionists have been quietly looking for an older chain of islands where the Galapagos creatures might have evolved. Their theory is that the older chain sank beneath the sea, forcing the animals to move to the Galapagos. Evolutionists have now announced that they have discovered the lost islands. They are east of the Galapagos Islands. These "islands" are today 6,500 feet beneath the waves. Samples dredged from them reveal what researchers called evidence of beach erosion. This shows, they say, that the older islands once extended above the waves. In truth, the dating of these islands, and the claim that the so-called "older" islands once stood above the waves, may easily be questioned. Nor can it be proven that the ancestors of the animals on the today's Galapagos once called these older islands home.

As Christians, we need to evaluate all of these guesses about the past in light of the established truth presented in the Bible.

Prayer: Father, I thank You that Your Word is trustworthy and that You have not left us to the shifting and uncertain wisdom of man. Help me to live my life in a way that shows everyone around me that Your Word is powerful and true. In Jesus' Name. Amen.

Ref: Malcolm W. Browne. 1992. "Galapagos' Diversity Came From Sunken Islands." *Stevens Point (Wis.) Journal*, Jan. 22.

Evolution's Influence

Matthew 26:41
"Watch and pray, that ye enter not into temptation: the spirit indeed is willing, but the flesh is weak."

Those who truly believe that the Bible is the Word of God consider its moral instruction to contain God's absolutes. Others think it is a book of ancient myths, while a third group considers its teachings generally good but not absolute.

A recent Gallup poll shows that most Americans say that religion is very important in their lives. The poll also shows that just half of Americans consider the Bible to be the Word of God. Only 16 percent said that they thought that the Bible was only ancient myths. And two-thirds of U.S. adults told pollsters that they think there are few, if any, moral absolutes. Right and wrong depend on the situation.

Those who have promoted evolution with the most enthusiasm have admitted that their goal has been to convince people that there are no moral absolutes. They have said that if people could be shown that everything could be explained without God, then people would stop thinking about moral accountability. Looking around us today, it could very well be concluded that the promoters of evolution are winning the battle for men's minds. Moral relativism is here and destroying lives and families, but history shows that this has happened many times before. Each time, Christians have turned the situation back to moral absolutes, and that can be done again.

Prayer: Dear Lord, I admit and confess that there have been times when I gave in to the temptation to think of my sin as "not as bad" as the sins of others. Forgive me, cover me with Your holiness, won for me on the cross, and fill me with a greater desire to do Your will, to Your glory. Amen.

Ref: "For Many, 'Situation Ethics' Are Replacing Moral Absolutes." *Washington Times*, Apr. 4, 1992.

Megakites

Luke 12:54-55
"And he said also to the people, When ye see a cloud rise out of the west, straightway ye say, There cometh a shower; and so it is. And when ye see the south wind blow, ye say, There will be heat; and it cometh to pass."

When we think of science and kites, we usually think of Ben Franklin flying his kite in the thunderstorm. Ben Franklin was not the first to use kites to study the weather. Nor was he the last. The U.S. Weather Bureau began using kites to study the weather after the invention of the box kite in 1893. Kites became important weather station tools for the next 30 years. As balloons were improved, the Weather Bureau's use of kites stopped.

In 1990, meteorologists again returned to the kite. Modern, high-tech materials gave meteorologists the hope of doing research with kites that was impossible with balloons. Modern meteorological kites can be "parked" miles high for weeks. A kite the size of two compact cars can weigh only a pound. Vents in the front of the wing-shaped kite inflate it so that it holds its wing shape. Instruments on the kite send down weather readings from various altitudes along the tether. The world record altitude reached by one of these kites is almost 12 miles!

The creation that God made for us has always inspired man's curiosity and wonder. Unfortunately, many of today's scientists would just as soon ignore God. The foundation upon which they build, however, is the scientific accomplishments of earlier scientists who, for the most part, believed in and worshiped our Creator.

Prayer: I thank You, dear Father, for the blessings we enjoy because of science. I ask that You would use the increased involvement of Your people in science so that science once again is used to glorify You instead of debate with You. In Jesus' Name. Amen.

Ref: Richard Monastersky. 1992. "Astride the Wind." *Science News*, Vol. 141, April 4, pp. 216-219.

Riders on the Wind

Mark 11:24
"Therefore I say unto you, What things soever ye desire, when ye pray, believe that ye receive them, and ye shall have them."

The tiny insect called iceplant scale has no wings. That's why scientists were surprised to learn that the creature uses sophisticated preflight and flight strategies to move from one iceplant to the next.

Iceplant scale infects iceplants, which are plants used in landscaping. To study the insect, scientists had to observe iceplant scale hatchlings under the microscope because they are only one-fiftieth of an inch long. It was in this setting that scientists saw an unexpected insect ballet.

Sensing a wind of ten miles per hour or more, a scale would determine the wind direction with its antennae and turn its back to the breeze. It then reared back on its hind legs and extended its antennae and legs. This doubles the insect's surface area and makes it possible for the insect to be lifted and carried by the wind. By extending its legs and antennae, the insect slows the rate of fall so that it can be carried further by the wind. Scientists noted that even insects that are one day old are knowledgeable about flight and ready to migrate.

These actions are designed to allow the wingless insect to take every advantage of the principles of flight. The only scientific conclusion that makes sense is that this knowledge was wired into the insect by its Creator, God Himself.

> **Prayer: Lord, I am amazed when I see the care and detail You have given to everything in the creation. Forgive me for the times I have treated You as though You were distant and I had been left here on my own. Amen.**

Ref: "Blown Away: Riders on the Wind." *Science News*, Mar. 31, 1984, p. 201.

A Matter of Life or Death

Genesis 3:2-3
"And the woman said unto the serpent, We may eat of the fruit of the trees of the garden: But of the fruit of the tree which is in the midst of the garden, God hath said, Ye shall not eat of it, neither shall ye touch it, lest ye die."

According to the Bible, God created the heavens and the Earth in a perfect condition, without disease or death. However, when man did his own will instead of God's will, death became a reality, just as God had warned. On the other hand, evolution says that death has nothing to do with man's disobedience to his Creator. Evolution also says that our genetic information has developed over the years to help us *survive*. Can these opposite claims be tested?

Biologists have discovered that living cells have a built-in suicide gene. The suicide gene remains dormant until it is necessary for the cell to die. For example, as a tadpole turns into a toad, its tail is no longer needed. As a result, the suicide genes in the tail cells instruct them to die.

In other words, living cells have a gene in them that prevents the death of the cell until its appointed time.

So in a sense, we can test the Bible's and evolution's conflicting claims about life and death. Why would evolution develop genes that order death? By definition, such a gene would not aid survival. However, God told our first parents that the day that they sinned, they would die. It was obviously the curse of sin that created the genetic instructions for cell death, on the very day that they sinned. What biologists are learning about the cell's workings fits much better with Scripture than with evolutionary claims.

Prayer: I thank You, dear Father, that You have provided a remedy and cure for my death through the forgiveness of my sins in Jesus Christ. By Your Holy Spirit help me to live the new life You have given me to Your glory. In Jesus' Name. Amen.

Butterfly Husbandry

Genesis 2:15
"And the LORD God took the man, and put him into the garden of Eden to dress it and to keep it."

The black madrone butterfly is one of the strangest butterflies in the world. It also has a highly unusual relationship with man.

The madrone butterfly inhabits only five species of trees in a small part of Mexico. In the caterpillar when they are ready to turn into butterflies, they spin tough bags of double-stranded silk. A single tree may have up to 20 silk bags hanging from it. The bags are so strong that the local Indians use them as bandages, containers, and flags.

Madrone pupae are rich in fat and protein. Since they are eaten by the local Indians, scientists became worried about the butterfly's future. The butterfly is already threatened by logging in the few forests it inhabits. Then scientists discovered that the Indians actually cultivate the butterfly. They regularly move the butterfly's silk bags from trees that have too many bags to trees with no bags, retying the bags to the trees. This care helps repopulate the butterflies in areas where they are scarce.

God has given us the Earth and its resources to use for our benefit and the benefit of others. He has also given us the intelligence necessary to care for the Earth as we use its resources. The Indians who tend the madrone butterflies show us by example proper management of the Earth's resources. Scientific knowledge is not as important as our attitude towards God's creation.

Prayer: I thank You, Lord, that You have so generously provided for all our needs. Help me to make efficient use of the resources You have provided to me, and help me not to be selfish and abusive with them. Amen.

Ref: Ron Cowen. "Butterflies in their Stomachs." *Science News*, Vol. 141, p. 236.

The King Bee

Romans 1:20
"For the invisible things of him from the creation of the world are clearly seen, being understood by the things that are made, even his eternal power and Godhead; so that they are without excuse:"

For over a century, it was a bee of legend. The native people on the Indonesian island where it lives call it the "king bee." The king bee was first described by a scientist in the 1800s. Then, despite a century of looking, the largest bee in the world was not seen again until the 1980s.

The king bee is about the size of a hummingbird. Scientists say that it sounds much like a hummingbird as it flies. Besides its size, the female king bee can be easily identified by its huge mandibles, or jaws. It uses the mandibles to collect resin that oozes from trees in its jungle home. The resin is collected, formed into a ball by the mandibles, and then held in them while being returned to the nest. Male king bees are smaller and have more proportional mandibles.

King bee nests are about the size of basketballs. They are built inside tree-dwelling termite nests. The bee uses tree resin to build its nest because the resin offers nothing of nutritional interest to the termites. The plastic-like tubes of tree resin within the bee nest cannot be destroyed by the termites. One of the scientists studying the bees reports that their sting is not as painful as a honey bee's and that, in general, king bees are not aggressive.

The huge king bee has been carefully designed to make its home safe, even among termites. The size and intelligence of this wonderful creature witness to the Godhead of our Creator.

Prayer: Lord, I thank You that I am filled with wonder as I see yet another example of the truth that there is no end to Your creativity and wisdom! Let me glorify You in my life by witnessing to Your saving truth in the forgiveness of sins. Amen.

Ref: J. A. Miller. 1984. "Jungle Jaws: Giant Bee Rediscovered on Indonesian Islands." *Science News*, May 12, p. 293.

Cooking Chemistry

Proverbs 1:7
"The fear of the LORD is the beginning of knowledge: but fools despise wisdom and instruction."

Want to make better souffle or meringue? Though many people never think about it, cooking involves chemistry. Good cooks have a knowledge of chemistry, even if they have never studied the subject in a classroom.

Beating egg whites for souffle or meringue is a tricky business. The egg whites must be beaten in just the right way to make a good foam. Even if you get a good foam, if the egg whites are beaten too much, the foam will be ruined. Scientists report that when beaten, the proteins in the egg whites form complex intricate molecules. More beating of the egg white begins to unravel these molecules and mixes them with air. The foam looks best, tastes best, and holds its form best when the proteins are only partially unraveled. Too much beating will unravel them too much.

Researchers also found that beating the eggs in a copper bowl will produce a much better beaten egg white, although it takes twice as long as in a glass bowl. They learned that the egg white absorbs a harmless number of copper atoms from the bowl, making the foam chemically more stable. If you don't have a copper bowl, a little cream of tartar will give the same effect.

We all do things every day that require knowledge of chemistry, physics and biology. This everyday scientific knowledge we all need shows us that science is a gift of God to all of us.

Prayer: Dear Father, I thank You for the practical knowledge of the world around me that I have never before thought of as science. Help me to use this knowledge in a way that helps others and is pleasing to You. In Jesus' Name. Amen.

Ref: "Copper: Whipping Egg Whites into Shape." *Science News*, Vol. 125, p. 296.

Smart Bacteria

Proverbs 10:14
"Wise men lay up knowledge: but the mouth of the foolish is near destruction."

How smart do bacteria need to be? You might be surprised. Scientists were.

Some bacteria swim toward things that they like and away from things they don't like. That may seem simple enough to us, but it's not that simple for a bacterium. *E. coli* swim through liquids using whip-like hairs called flagella. Because they are so tiny, their movement through water is like a human trying to swim through molasses. They only move when they rotate their flagella. When rotated counterclockwise, their six or eight flagella wind together, forming a propeller that rotates at 140 revolutions per second! This moves the bacterium along in a straight line. However, periodically the bacterium will reverse its propeller, causing the flagella to unwrap and spinning the bacterium into a tumble. This led scientists to wonder how a bacterium ever got where it wanted to go.

Scientists also realized that for a bacterium to get to something that attracts it, it must have sensors and a memory. A bacterium must be able to say to itself, "It smells better here than where I was a moment ago. I'll keep moving in this direction." In other words, scientists are trying to figure out what makes bacteria so smart and gives them so many unexpected abilities.

The intelligence and abilities of bacteria are no mystery to those who know that we have an intelligent Creator. We would expect everything He made to reflect His great abilities.

Prayer: Father in heaven, for Jesus' sake, forgive me for not reflecting the intelligence You have given me. Help me to make better use of my intellect for You and for others. In Jesus' Name. Amen.

Ref: Dietrick E. Thomsen. 1984. "Swimming for the Good Life." *Science News*, Vol. 125. May 12, p. 298.

Crystalline Silk

2 Corinthians 4:7
"But we have this treasure in earthen vessels, that the excellency of the power may be of God, and not of us."

Scientists have been studying the safety line used by the common garden spider to save itself when falling. This safety line is as strong as nylon and has twice as much stretch.

The drag line silk is only one of several kinds that the spider can make. Scientists report that the silk is made of sections that are crystalline and sections that are like rubber. This makes it both stretchy and strong. Amazed chemists note that usually when you make something stronger, you also make it more likely to break rather than stretch. Since spider silk is twice as strong as silkworm silk, why hasn't it been used to make more things?

To harvest enough spider silk for study, scientists place a spider on a piece of cardboard. Then they bump the spider off the cardboard. As the spider makes the drag line silk, a scientist turns the cardboard to reel it in. One spider made half-a-mile of silk in ten minutes! This fact might give you a hint about why spider silk isn't used to manufacture items. It's too fine. In the early eighteenth century, an inventor did make gloves and stockings from spider silk. However, his experience showed that it would take more than half a million spiders to make a pound of silk!

The spider and its silk are engineering marvels that clearly show why we must rely on a Creator to explain the world in which we live.

Prayer: Lord, I thank You for the intricate wonders that surround me in Your creation. As I do my tasks, help me to put my best workmanship into them. Help me to do everything as my best offering to You. Amen.

Ref: J. A. Miller. 1984. "Spider Silk Stretch and Strength." *Science News*, June 23, p. 391.

Intelligent Artists

Psalm 90:17
"And let the beauty of the LORD our God be upon us: and establish thou the work of our hands upon us; yea, the work of our hands establish thou it."

The ancient, yet beautiful, paintings in the Lascaux cave in France were discovered nearly 70 years ago. The paintings are thousands of years old, and there seems to be no other record of the painters. In those 70 years, researchers studying the paintings have deepened their admiration for the intelligence and skill of the artists.

Researchers have found the scaffolding used by the artists as well as the stone lamps they used to light the work area. The palettes and even the pigments still on the palettes have also been found.

Using scanning electron microscopes and other high-tech equipment, researchers learned that the artists collected a range of local minerals. Mineral deposits within 15 kilometers of the cave were able to provide a full range of colors. The minerals were ground into fine powder and mixed to create pigmented paints that hold their color and beauty to this day.

As we look around the creation, we can clearly see that our Creator has a love for beauty. When He made us, He gave that appreciation for beauty to us as well. In our day, we have also seen how man's rebellion against God can twist art into ugliness and evil. Christians today need to face the question of what is art and what isn't. As we do this, we can ask ourselves whether a given piece of art reflects our Creator's love of beauty or whether it reflects the ugliness of man's sin and rebellion against his Creator.

Prayer: Lord, I thank You for the beauty that surrounds us in Your creation. Increase the appreciation for true art that reflects Your love of beauty among Your people, and deliver us from art that reflects the ugliness of man's sin. Amen.

Ref: Ivars Peterson. "Pigment Processing for Cave Paintings." *Science News*, Vol. 125, p. 348.

Moth Talk

1 Corinthians 15:28
"And when all things shall be subdued unto him, then shall the Son also himself be subject unto him that put all things under him, that God may be all in all."

Scientists are learning that there is much more going on among those moths that flutter around your yard light than they ever expected.

Scientists have long known that moths communicate with each other using hormones called pheromones. A female moth who is ready to mate will send a pheromone into the air. Males who sense the pheromone will seek out the female. Research now shows that the moths' system is much more elegant than this.

The female who wants to contact a male doesn't simply send out a scented invitation. Rather, she releases her pheromones in a series of pulses. These pulses are released at the rate of from one to more than two pulses per second. Scientists used powder to study the pulses. To their surprise, they discovered that the pulses can be propelled for up to a meter. That's equivalent to you being able to blow out a candle from 60 feet away! These pulses had researchers puzzled. Then they realized that the pulses would help male moths locate a female as they were drawn near by the scent of the pheromone. This conclusion seems to be confirmed by the fact that the male's antennae send a signal to its brain that has the same frequency as the female's pulses.

The creation shows much more than design. Each of the details of the creation have the elegant intricacy of a master Designer. This is, in part, what St. Paul refers to when he writes that the very Godhead of the Creator can be seen in the creation.

Prayer: Father in heaven, I admire and thank You for the excellence of Your workmanship in the creation. As I do my daily tasks, help me to take care to put the best workmanship into my work so that You will be glorified in everything I do. In Jesus' Name. Amen.

Ref: "Smoke Puffs for Sexual Communication." *Science News*, June 23, 1984, p. 393.

Jungle Fungus Fighter

Matthew 16:26
"For what is man profited, if he shall gain the whole world, and lose his own soul? or what shall a man give in exchange for his soul?"

Throughout the open markets of Kenya, shopkeepers sell leaves and strips of bark from the warburgia bush. The leaves and bark are used to relieve toothaches, fevers, constipation, and the taste of meat gone slightly sour. Medical researchers have been examining extracts from the bush for use with antibiotics.

The immune system of a healthy person will keep fungal infections under control quite easily. When the immune system has been weakened by illness, however, heavy doses of antibiotics, chemotherapy, and fungal infections can be life-threatening.

Researchers discovered that two extracts from the warburgia bush can boost the effects of antibiotics. One of the extracts, coupled with an antibiotic used against yeast cells, resulted in a treatment that was 16 times more powerful than the antibiotic alone. Warburgia extracts also boosted antibiotics' effects against various fungal infections, including the fungus that causes athlete's foot.

How many medicines has our Creator hidden in the plants of the Earth for us to discover? We can thank Him that, in His mercy, He has provided so many earthly remedies for the disease that has resulted because of sin. What is more important, however, is that we can thank Him for the forgiveness of sins that was won by Christ for us in His suffering and death.

Prayer: Thank You, Father, for Your generous gifts that have helped us reverse some of the earthly effects of sin. Bless our researchers as they continue to search for more medicines. However, help me always remember to look to Christ for the ultimate permanent solution to my sin. In His Name. Amen.

Ref: D. Franklin. 1984. "Plant Extract Fights Fungus." *Science News*, June 16, p. 375.

Evolutionary Medical Ethics

Ezekiel 34:4
"The diseased have ye not strengthened, neither have ye healed that which was sick, neither have ye bound up that which was broken, neither have ye brought again that which was driven away, neither have ye sought that which was lost; but with force and with cruelty ye have ruled them."

A professor of medical ethics used an old evolutionary myth to support the removal of organs from a baby born without a brain. The girl's parents wanted to donate her organs for transplant while she was still alive so that the organs would be healthy and usable. A circuit court judge, however, noting that the child was still alive, refused to allow the taking of any of the child's organs vital for life.

In support of taking the child's organs, Yale professor of medical ethics Dr. Robert J. Levine said, "Our brain stems do not differ substantially from the brain stem of a fish." He added that the child has "more in common with a fish than a person." These comments reflect a discredited evolutionary theory. The theory holds that in passing from fertilized egg to birth, the child goes through the evolutionary steps that led up to humans.

Did the infant have more in common with fish than humans? While the unfortunate infant had only a brain stem, every bit of genetic information in the tissue was fully human. That brain tissue operated human organs within a human body. No fish brain in the world has either of those characteristics.

Does what you believe about origins matter? Fraudulent evolutionary science leads us to a steep and slippery slope that cheapens human life, which has been created by God.

Prayer: I pray, Dear Father, that You would look upon our society and its treatment of its weakest members with gracious eyes. Equip Your people with the wisdom and resolve necessary to turn our nation back to the road that protects the lives of its most helpless members. In Jesus' Name. Amen.

Ref: "Parents Fought to Donate Organs of Anencephalic Newborn." *Minneapolis Citizens Concerned for Life Newsletter*, April, 1992, p. 2.

The Queen of All Herbs

Jeremiah 46:11
"Go up into Gilead, and take balm, O virgin, the daughter of Egypt: in vain shalt thou use many medicines; for thou shalt not be cured."

The ancient Greeks called it the "queen of all herbs." The earliest mention of this common shrub refers to its medicinal value.

The peonies with which most of us are familiar are painstakingly developed hybrids. Greece, as well as much of the rest of the world's temperate zone, offers many species of wild peony. It gets its name from the Greek god Paeon who was, in legend, a follower of the Greek god of medicine. According to the myth, Paeon used the healing properties of the peony to heal Pluto after he was injured in the Trojan War. This made the Greek god of medicine jealous and he wanted to kill Paeon. To save Paeon's life, says the legend, Pluto changed Paeon into a peony plant.

According to medicinal encyclopedias dating back as far as the first century A.D., the peony was thought to offer a cure for lunacy and epilepsy. Peony roots were said to offer medication for kidney and bladder problems, and abdominal pains. The seeds were said to be effective against nightmares and hysteria. Though still used medically today in the Greek countryside, none of these applications have ever been investigated or confirmed by modern science.

Medical research may someday confirm that the ancients understood more about the medicinal effects of the peony than science does today. Is it possible that such knowledge was originally given by God to Adam and Eve?

Prayer: I thank You, Lord, for the blessings You have brought us through modern medicine. I also thank You for the medicines You have given us through the plant and animal kingdoms. But most of all, I thank You for the eternal remedy to sin we have in the forgiveness of sins through You. Amen.

Ref: Julie Ann Miller. 1984. "Greek Portraits of a Queen." *Science News*, Vol. 126, July 28, pp. 56-57.

A Universal Graft?

Genesis 9:3
"Every moving thing that liveth shall be meat for you; even as the green herb have I given you all things."

Many uses have been found for the intestines of pigs. Some astonishing test grafts now suggest that a portion of the pig's intestine may provide doctors with an almost universal graft. Tests show that the grafted intestine can be used for several types of repairs and offers no problems with rejection.

Pig intestine is made up of three layers of tissue. The outer layer is muscular tissue. The inner layer is the tissue involved in digestion. They are held together by a layer of mostly collagen connective tissue. Purdue University veterinarians have developed a process for separating the collagen layer from the other two layers.

This layer, called SIS, resists blood clots. For this reason, researchers first tested it as a possible replacement material for damaged aortas. It worked well. And to their surprise, they found that after two months, the SIS had been dissolved by the body and replaced by a new, undamaged aorta. The new aortas remained strong and healthy five years after the surgery. Researchers also investigated the ability of SIS to rebuild knee ligaments and Achilles tendons. Weeks after surgery, the SIS had become fully developed ligament and tendon tissue.

This research reminds us that God has created plants and animals to help us, as well as for us to tend and care for.

Prayer: Dear Father, I thank You for the wonderful design of Your creation that places so many different kinds of living things together within so much harmony. As we learn more about using these blessings, help us also to develop our knowledge and skill so that we take more responsible care of the creation. In Jesus' Name. Amen.

Ref: C. Ezzell. 1992. "Pig Intestine Yields Versatile Tissue Graft." *Science News*, Vol. 141, April 8, p. 246.

Green Toothpaste?

Genesis 1:31
"And God saw every thing that he had made, and, behold, it was very good. And the evening and the morning were the sixth day."

As a child, Isao Kubo's grandmother always told him to drink green tea after he ate something sweet. Today, as a University of California chemist, Kubo has shown that his grandmother's advice was scientifically sound.

It's been known for about ten years that tea can help fight dental cavities. A substance in the tea helps prevent the bacteria that cause cavities from latching onto the surface of a tooth. But researchers found that this effect was too small to make tea as effective as it is known to be in preventing cavities. This led scientists to continue their search for something else in tea that makes it so effective.

Kubo discovered that a class of molecules found in abundance in green tea actually kill the bacteria that cause cavities. Tests show that these natural chemicals, while harmless to us, are also able to kill at least two different kinds of molds, three yeasts, and eight kinds of bacteria. It is effective against some of the bacteria that cause gastrointestinal disease and acne. However, you don't have to drink green tea to take advantage of this powerful, natural cavity fighter. It is also found in coriander, sage, thyme, and is added to some ice creams, candy, chewing gum, and baked goods. Because of this research, we may someday find it in toothpaste.

One wonders whether, in the perfection that existed before sin, the creation was filled with many additional natural substances that prevented the health problems we experience today.

Prayer: Father in heaven, I thank You for the work of those who are discovering the natural substances You have created to help keep us healthy. I pray that You would not allow my concerns about physical health make me lose sight of my need for spiritual health through the forgiveness of sins that was won for me by Your Son, Jesus Christ, on the cross. Amen.

Ref: Janet Raloff. 1992. "Another Reason to Drink Green Tea." *Science News*, April 18, p. 253.

A Fish Story

Genesis 1:20
"And God said, Let the waters bring forth abundantly the moving creature that hath life, and fowl that may fly above the earth in the open firmament of heaven."

School biology textbooks offer students good facts but often bad interpretation of those facts. For example, when explaining how fish supposedly evolved into air-breathing land animals, students are usually offered a great deal of misinformation about living and extinct fish.

It is not true that lungfish are an odd class of fish and that no other fish have lungs – many do and some have lungs as well as gills. The textbook explanation says that all early fish had air-filled swim bladders, as most fish do today. The textbooks go on to say that the bladders in some of these fish eventually evolved into lungs. Then, as these fish spent more time on land, some of them gradually lost their gills and developed from amphibians into reptiles. The facts are shown in the fossil record where, according to evolutionist's own inflated system of dating, lungs actually appeared in fish before swim bladders! A logical evolutionist would have to conclude that the fish's air bladder evolved from lungs. The problem for evolutionists is that this is precisely opposite of what they need for the theory. If we telescope the evolutionary years into historical years, however, the fossil record agrees with Scripture. In other words, fish with lungs and fish with swim-bladders both appeared on day five of creation.

The lesson for us here is that when science tells the truth, it actually confirms those seemingly simple statements found in the Bible.

Prayer: Dear Lord, I thank You that among the din of man's mistaken opinions, I have trustworthy knowledge available in Your Word. Increase my resolve to learn and apply Your Word so that I am not so easily misled by man's mistaken opinions. Amen.

Ref: Joachim Vetter. "Something fishy about lungs." *Creation ExNihilo*, Vol. 14 No. 1, pp. 46-47.

Octopus School

Philippians 2:5-7
"Let this mind be in you, which was also in Christ Jesus: Who, being in the form of God, thought it not robbery to be equal with God: But made himself of no reputation, and took upon him the form of a servant, and was made in the likeness of men."

Can an octopus be as smart as a dolphin or even a human? The octopus is thought by those who believe in evolution to be an early, primitive form of life. They certainly look primitive, but that's a subjective opinion based upon the many imaginative illustrations we have all seen in books and films. And we might also say that they do not appear to be very intelligent, but again, do we really know for sure?

In some recent work, researchers had two tanks of octopi and placed a red ball in one and a white ball in the other. The octopi in each tank were trained to attack the balls in their tank. They then placed another tank of untrained octopi close by where they could watch their trained companions. A red and a white ball were then dropped into their tank and, sure enough, those who had watched a red ball being attacked did the same thing but ignored the white ball. Not only did the untrained octopi attack the same color they had watched being attacked, but they learned the lesson much faster than the first group.

This discovery about the intelligence of the octopus is even more surprising because they're not social creatures. Social vertebrates tend to appear more intelligent because they learn from one another. In a similar way, the Creator of this universe became a human being in the person of Jesus Christ so that we could learn by watching and listening to Him.

__Prayer: I thank You, Lord Jesus, for becoming like us so that we could more easily understand Your message for us. I especially thank You that You endured the suffering of the cross so that my sins might be forgiven and I might be restored to God. Amen.__

Ref: M. Stron. "In the lab, it's octopus see, octopus do." *Science News*, Vol. 141, p. 262.

Are We Making Progress?

Genesis 2:16-17
"And the LORD God commanded the man, saying, Of every tree of the garden thou mayest freely eat: But of the tree of the knowledge of good and evil, thou shalt not eat of it: for in the day that thou eatest thereof thou shalt surely die."

It would seem almost ludicrous to ask the question, has mankind made progress since his appearance on this planet? Many would answer that it should be perfectly obvious that man has come a long way from primitive beginnings in the cave.

Until recently, historians have always spoken of the idea of progress in history, but this is now being seriously questioned. While it is true that there have been improvements and innovations in technology throughout human history, each step forward has brought with it a mixed blessing. For example, the movable-type printing press introduced about 1447 reduced the cost of printing, thus making, say, Bibles much more readily available. While this helped to establish moral absolutes, those same printing presses also produced pornography and politically dangerous ideas from the Greeks. More recently, television seemed like free entertainment, but overindulgence has robbed many of their health and the ability to think for themselves.

In those very early chapters of the Bible we are told how in disobedience to God's command, our first parents, Adam and Eve, ate the fruit of the tree of the knowledge of good and evil. It is mankind's fallen nature that has caused the good and the evil to come together as a package. We have made progress in technology, but we have also regressed morally. As it was in the days of Noah, the Earth has become filled with war and violence by man's use of good inventions put to evil purpose.

Prayer: Dear Lord, we thank You for all the inventions and advances in technology that You have given to help us in our continuing fall from perfection. May we be mindful of these gifts and always use them for good and not evil.

Ref: Nisbet, Robert. *History of the Idea of Progress.* New York: Basic Books, 1980.

Insulated Blackbirds

Job 12:7, 9, 10
"But now ask the beasts, and they shall teach thee; and the fowls of the air, and they shall tell thee...Who knoweth not in all these that the hand of the LORD hath wrought this? In whose hand is the soul of every living thing, and the breath of all mankind."

European blackbirds love southern Germany so much that many of them stay for the winter. Some of the blackbirds migrate to North Africa, Spain, or southern France to escape the cold. Like the robin in North America, however, some European blackbirds don't migrate.

European blackbirds are able to live through the cold German winter because they are designed with several features that use sophisticated principles from biochemistry and physics. To cope with the winter cold, the birds fluff up their feathers. This creates more insulating dead air spaces between their warm bodies and the cold. Then they tuck their head inside the feathers, making their body into a ball. A round ball is the most heat-conserving shape possible for any object. The ball also protects poorly insulated parts of their bodies, like feet, legs, and beak.

Food is also more scarce in the winter. On top of that, the birds' need for energy is five times higher at 20 degrees below zero Fahrenheit than it is during the summer. The blackbirds reduce their energy needs by lowering their body temperature at night when they are curled into a ball. Scientists have found that, as a result, European blackbirds are in no danger of freezing, even at 20 degrees below, as long as they can get enough food!

The intelligent special designs found in the European blackbird reflect a sophisticated knowledge of biochemistry and physics. By its very existence, the European blackbird glorifies our Creator God.

Prayer: I thank You, Lord, that You have designed Your creation not only for the benefit of Your creatures, but also so that it is nearly impossible to attribute Your handiwork to someone or something else. Amen.

Ref: "Stay-at-home blackbirds cope with cold." *Science News,* Aug. 25, 1984, p. 119.

One Giant Step for Mankind

Genesis 1:27
"So God created man in his own image, in the image of God created he him, male and female created he them."

The lead article and front cover picture of the July 23, 2001, issue of *Time* magazine promised its readers that a recent discovery in Ethiopia has brought science tantalizingly close to solving the puzzle: "How Apes Became Human."

The discovery consisted of eleven small pieces of fossil bone, including a jawbone and some teeth, from at least five different individuals. The reader is told that these fragments were found scattered over several sites and, fortunately, were in the same rock layer. That layer was dated at about five and a half million years old. This discovery alone would not have compelled the *Time* editors to even report it. One of the precious bones, however, was a toe bone whose features "proves the creature walked on two legs." Apes do not habitually walk on two legs and, thus, it was this toe bone that quickly moved the discovery from relative obscurity to the front page of *Time* magazine. The article hails these bone fragments as the earliest link between the ape and man.

Near the end of the article is a revealing statement: The toe bone was not only separated in time by several hundred thousand evolutionary years but was also found ten miles away from the rest of the bone fragments! *Time* editors may believe that this is all good evidence that man evolved from the ape's image, but Creation Moments suggests it is more credible to believe that man was created in God's image.

Prayer: Heavenly Father, You have told us that the world's wisdom is plain foolishness. Please help us to stay close to Your Word, to believe it, to love You and never be deceived by the world's wise men. In Jesus' Name. Amen.

Ref: Michael D. Lemonick & Andrea Dorfinan. "One Giant Step for Mankind." *Toronto: Time (Canadian ed.)* July 23, 2001, Vol. 158, pp. 48-55.

The Problem of Genius

Genesis 5:1
"This is the book of the generations of Adam. In the day that God created man, in the likeness of God made he him."

We have for so long been led to believe that early man evolved from the animal kingdom and, thus, began with little intelligence. But is our mental capacity today really the result of evolution from brute beginnings?

From time to time individuals display a mental capacity that far exceeds their need of it. This anomaly presents a serious challenge to the theory of evolution. Consider the example of George Koltanowski, International Chess Grandmaster. In 1960 he played against 56 other chessmasters simultaneously, winning 50 games and tying 6. He took about 10 seconds per move and was blindfolded throughout the entire nine-hour marathon! This is an incredible achievement of the human memory. Evolution is said to depend upon chance mutations that give the individual a survival advantage, but it is difficult to see how this would explain George Koltanowski's extraordinary memory capacity.

It is sometimes suggested that most of us only use part of our brain capacity while a genius uses almost all of it. But why would an evolutionary process have given us all this capacity in the first place? A far more reasonable argument is that today's genius is simply an unusual retention of ancestral brain capacity that reminds us that God created man in His own image.

Prayer: Dear Lord, please help us to recognize that You created us in the beginning in Your perfect image. Sin has caused us to fall a long way from that perfection, but You have provided a way through Jesus to make us whole again. In Jesus' Name. Amen.

Ref: Hooper, J.R. & A. Whyld. *The Oxford Companion to Chess.* Oxford University Press 1992, p. 206.

New Light on Television Violence

Matthew 15:19
"For out of the heart proceed evil thoughts, murders, adulteries, fornications, thefts, false witness, blasphemies:"

Scripture warns us that how we occupy our minds affects our behavior. Unfortunately, modern voices have said something else. They oppose the public expression of Christianity and decency because they say that it may have a negative effect upon people. At the same time, they tell us that when children study New Age religions, homosexuality and situation ethics at school or see violence on television, these influences will not have a negative effect upon them!

In recent years, parents have been subjected to conflicting "scientific findings" about the effect of television violence on children. Most Christian parents have had a clear sense that television violence has a bad effect on their children even if their children don't show any effects right after watching a violent program. New research clears up this seeming contradiction and supports what our Lord taught us.

Researchers from the University of Illinois showed that children who are already violent don't show much increase in their violent actions after watching violent television. Children who are not normally violent and who watch violence on television do not show much increase in violent behavior either, unless they are faced with a frustrating situation. Then, instead of reacting with little or no violence, they react with a much greater degree of violent behavior.

As the Lord who created us said, what we put into our minds will be reflected in our behavior.

Prayer: Dear Lord, I thank You for Your wholesome Word. Help me to read and study it in faith so that my mind may be filled with Your will. Yet, let this not be my hope, but let me hope in the forgiveness of my sins that You won for me on the cross. Amen.

Ref: B. Bower. "Kids' Aggressive Behavior Linked to Watching TV Violence." *Science News*, Vol. 126, p. 190.

Miracle Bugs

Job 15:8
"Hast thou heard the secret of God? And dost thou restrain wisdom to thyself?"

God designed and made the universe from nothing. That's creation. Man takes what God has made and simply rearranges it to make things that help him. That's building. The wonderful abilities that God built into plants and animals come from the genetic information that God built into them. Man has now started to use the genetic information created by God to build harmless creatures that better serve our needs.

For example, the silk of the golden orb spider is five times stronger than steel, yet more elastic than any similar material we could manufacture. It could be used to make light and comfortable bulletproof vests and helmets. The problem is, it takes over half a million spiders to make a pound of silk. So scientists have taken the spider gene that makes the silk and put it into a harmless bacterium. When that bacterium divides, it makes more bacteria, complete with the gene. Today, there are millions of bacteria making usable amounts of spider silk.

This technology has resulted in what has been called the greatest revolution since industrialization. Bacteria are now making insulin, vaccine, and even chocolate. Today, bacteria are mainly used to manufacture medicines.

Man has learned a great deal about how to make use of the creation. However, the genetic information that makes this all possible in the first place is a product of the genius of our Creator.

Prayer: Dear Father in heaven, I thank You that You have given us the ability to learn how to make productive use of what You have created. I pray that man will not misuse this technology. In Jesus' Name. Amen.

Ref: Doug Stewart. "These Germs Work Wonders." *Readers Digest*, pp. 83-86.

A Sting Operation on Potato Beetles

Psalm 89:12a, 14
"The north and the south, thou hast created them...justice and judgment are the habitation of thy throne: mercy and truth shall go before thy face."

The potato beetle is one of the most noxious pests that can attack a farmer's crop. It is resistant to insecticides. As a result, it has been America's most effective destroyer of potato, tomato, and eggplant crops. A tiny surgeon from South America, however, has shown promise in combating the potato beetle.

The South American wasp is only about the size of a gnat. The wasp uses a specialized stinger, something like a mosquito's, to drill up to 50 holes in potato beetle eggs. While the wasp will occasionally eat the egg's contents, its true goal is far more sinister. The wasp is preparing the beetle egg so that it can insert one of its own eggs within the larger beetle egg. As a result, more than 80 percent of the beetle eggs studied hatch out into South American wasps. And those wasps are looking for more beetle eggs.

Researchers were tipped off to the wasp's possible interest in potato beetles by its habit of attacking a South American cousin of the beetle. They say that the wasp does not harm other plants, nor does it have an interest in the eggs of other insects. The problem, however, is that it is difficult to establish the South American wasp in North America because it cannot survive the more severe winters.

The fact that the created world is balanced with creatures that can help *keep* other creatures under control when they become destructive is yet another evidence of the Creator's design.

> **Prayer: I thank You, dear Father in heaven, that You have made us able to learn about the balances You have designed into the creation. I pray that these balances may be helpful in convincing scientists that You are the Creator so that they might desire the relationship with You that is available through Jesus Christ. Amen.**

Ref: J. Raloff. 1984. "The Sting: Wasps Use Pets Eggs." *Science News*, Sept. 8, p. 149.

An Ancient Industrial Secret

Numbers 15:37-38
"And the LORD spake unto Moses, saying, Speak unto the children of Israel, and bid them that they make them fringes in the borders of their garments throughout their generations, and that they put upon the fringe of the borders a ribband of blue:"

Through Moses, God commanded the Israelites to dye the tassels on the corners of their garments blue. The secret to making this blue dye appears to have been lost about 500 years after the birth of Christ.

The Hebrew description of the blue that the Israelites were to use on their garments refers to what we today call hyacinthine purple. Researchers searched for the original process for making the dye, using clues from the Bible, the Talmud and ancient trade records. Ancient references to the dye go back at least 3,600 years. While the ancients knew how to make this dye, we moderns don't.

Once researchers identified the correct color, they began to search for its origin. The Talmud says that the dye must be made from the uncontaminated extract from a shellfish. Eventually, scientists discovered a species of shellfish, the murex, that produces the raw material for exactly the color described in the Bible. Their final challenge was to perfect the dyeing method, which one researcher called one of the great industrial secrets of the ancient world.

In our modern technological age, it's good to be reminded that we are not the only generation to develop sophisticated industrial secrets. This broader perspective helps us to understand claims that ancient man was not so primitive as some today might imagine.

Prayer: Lord, I thank You for the blessings that are ours today because of technological and industrial development. I ask that You would use discoveries such as this blue dye to help us understand that ancient man was just as clever and intelligent as we are today. Amen.

Ref: D. Franklin. "Blue-purple Dye of Antiquity Reborn." *Science News*, Vol. 126, p. 148.

Crunchy Medicine

1 Corinthians 15:49
"And as we have borne the image of the earthly, we shall also bear the image of the heavenly."

Celery is an old Asian folk remedy for mild cases of high blood pressure. In one unscientific test, the father of a University of Chicago medical student lowered his mild hypertension simply by adding a quarter pound of celery to his diet every day for a week.

Now his son, a doctor named Quang T. Le, believes he has discovered how celery lowers blood pressure. His discovery could change the way high blood pressure is treated. Hypertension is caused when the smooth muscle lining of the blood vessels tightens. This makes vessels narrower, making it more difficult for blood to get through. Stress hormones in the blood cause the muscles to tighten.

Dr. Le discovered that celery contains a chemical that relaxes the smooth muscle tissue of the vessels. It does this by getting to the root of the problem. Stress hormones are made by the body using a specific enzyme. The natural chemical in celery prevents the enzyme from forming into stress hormones, thus preventing the chemistry that creates high blood pressure. The researchers hope that the chemical will lead to a more effective treatment for high blood pressure. They also recommend against eating celery to lower blood pressure since it contains sodium and other chemicals that can be harmful to some people when taken in high dosages.

How did the ancient Asians learn about this natural solution to hypertension? Clearly, so-called ancient man had a great deal of knowledge about the creation that we are only just learning.

Prayer: Lord, when people today talk about previous generations as simple and primitive, they are demeaning Your workmanship, for man was made in Your image. Through the forgiveness of my sins, help me live in Your image today. Amen.

Ref: Carol Ezzell. 1992. "Celery Studies Yield Blood Pressure Boon." *Science News*, May 9, p. 319.

A Star Takes a Swan Dive

Romans 8:20-21
"For the creature was made subject to vanity, not willingly, but by reason of him who hath subjected the same in hope, Because the creature itself also shall be delivered from the bondage of corruption into the glorious liberty of the children of God."

The life cycle of a star is supposed to be millions of years; thus, human observers should not be able to see the birth and death of an individual star. Now a star that first became visible in the year 1600 is causing some astronomers to rethink their theories about the time it takes for stars to evolve.

According to the evolutionary way of dating things, some stars, like our sun, have been burning for billions of years. Astronomers say that stars go through different stages in their lifetimes. As a result, they can tell by color and brightness just how old a star is. The Bible says, of course, that the stars were created by God in the much more recent past.

In the year 1600 a star in the constellation of Cygnus, the swan, suddenly became bright enough to see from Earth. Astronomers report that the star called P Cygni has gotten gradually brighter and is half again as bright today as it was in 1700. This brightening is expected in the normal lifetime of a star. The problem is, at this rate the star is aging too fast, even for a large star. One scientist noted that the time scale for star aging can be measured in centuries, rather than hundreds of centuries. In other words, so-called "old stars" can be much younger than expected. That being the case, God's universe itself is seen to be much younger than expected – perhaps only thousands and not billions of years old.

Prayer: *Dear Father, as I am reminded that the entire creation is growing old and wearing out because of sin, cheer me again with the sure knowledge of the forgiveness of sin won for me by Your Son, Jesus Christ, on the cross. In His Name. Amen.*

Ref: Ron Cowen. 1992. "Astronomers Watch a Star Age." *Science News,* Vol. 141. May 2, pp. 298-299.

The Miracle of Hearing

2 Samuel 22:50
"Therefore I will give thanks unto thee, 0 LORD, among the heathen, and I will sing praises unto thy name."

Your high school biology textbook may have given you the idea that modern science understands how hearing works. While science understands the general principles, however, researchers are still trying to understand the details.

Hearing starts with the skin and cartilage on the outside of our heads that we call the ear. This tissue is carefully designed to collect sound waves and focus them into the hole in the lower center part of the external ear. That hole is a tube that runs for about an inch to the eardrum. As the eardrum vibrates with the incoming sound, tiny specialized bones pick up the vibrations. One of these bones changes the vibrations into hydraulic pressure.

The sound, now converted to hydraulic pressure, is sent to the cochlea. This coiled, bony canal is lined with tissue that has four long rows of hair cells. A vibration as small as the width of an atom will move these hairs. Each movement of the hair causes a change in the electrical potential of the hair cell's membrane. This change triggers a current that is fed to the auditory nerve and interpreted as sound by the brain. Each of the hairs in the ear responds best to a specific frequency.

The ability to hear could have been engineered using a much less complex design. Such a design, however, would leave us unable to hear the subtleties of an orchestra or a bird's song. Our Creator is not only capable of excellent workmanship, but also generous in His creation.

Prayer: I thank You, dear Father, for the richness of the sounds in Your creation. Let my voice praise You and tell others of Your wondrous works, especially the salvation You have prepared through the forgiveness of sins through Your Son, Jesus Christ. Amen.

Ref: Deborah Franklin. 1984. "Crafting Sound from Silence." *Science News,* Vol. 126, Oct. 20, pp. 252-254.

Plants That See

Acts 14:17
"Nevertheless he left not himself without witness, in that he did good, and gave us rain from heaven, and fruitful seasons, filling our hearts with food and gladness."

We are all familiar with the fact that plants can sense light. They depend upon light to make their living, so it's logical that plants have been designed with the ability to sense light. If plants could move, would they be able to see?

Green algae are classified as plants. And strangely enough, there are algae that can swim and see. To their surprise, scientists have discovered several species of marine and freshwater algae have eyespots. Each has a patch of membrane that has about 100,000 pigment molecules. As the algae swim, they react to objects in the water very differently than algae that don't have vision. The researchers' greatest surprise came when they discovered that the chemical pigment responsible for area vision is the same chemical that makes our vision work.

There is an interesting sidebar to this story. The pigment, called rhodopsin, is surprisingly similar to that in cows, in man and in the alga that can see. Researchers were surprised to find that this same chemical that makes our vision possible is also used by such a supposedly "simple" creature. This is not at all what their evolutionary theories would lead them to expect.

Our Creator is wise. When He wanted to give the ability to see to His creatures, He used the same basic principle whether He was making algae, cows, or humans. The fact that He did not use a simpler method to make so-called "simpler" creatures see is yet another way in which His handiwork witnesses against evolution.

Prayer: *Dear Lord, I thank You that all of the creation, from highest to lowest, glorifies You by witnessing to the genius of Your creating work. Help my life be to others a visible witness of the new life I have through You. Amen.*

Ref: J. A. Miller. 1984. "Eyespot for an eye: Algae and Animals share visual pigment." *Science News*, Nov. 10, p. 295.

Plastic Farms

Genesis 1:28
"And God blessed them, and God said unto them, Be fruitful, and multiply, and replenish the earth, and subdue it: and have dominion over the fish of the sea, and over the fowl of the air, and over every living thing that moveth upon the earth."

Have you ever been to a farm that grows plastic? In just a few years this scenario may become a common reality. Most plastics today are made from petroleum. Plastics made from petroleum are not considered biodegradable. Researchers have been searching for a way to get plants to produce plastic resin. Such plant-produced resin could be used to make biodegradable plastics. The few biodegradable plastics now on the market are made by mixing starch or other plant material with petroleum products.

Researchers from James Madison University in Harrisonburg, Virginia, have announced that they have engineered a plant that makes polymer resin. The resin is a biodegradable form of the petroleum-based plastic polypropylene that's now used to make containers and wrappers. While the plant-produced polypropylene looks and acts like its petroleum-based cousin, it is easily broken down by enzymes in the soil. To produce the plastic, researchers added to a relative of the rape plant a gene that enables a common soil bacterium to make the plastic. They say that preliminary studies indicate that plants can be engineered to make a variety of plastics. Further research is being done to increase plastic production and design a plant that is easy to grow on a large scale.

When God placed man on Earth, He told us to make use of it for our good and the good of the rest of the creation. Our discovery and use of the abilities God placed into the creation fulfills this command and improves life for all of us.

Prayer: Dear Father in heaven, I thank You for all the abilities You have given to the plants and animals You created. Help me to use the abilities You have given me to glorify You and witness to the salvation You have made available to us in Jesus Christ. In His Name. Amen.

Ref: Amal Kumar Naj. "Plant's Genes are Engineered to Yield Plastic." *The Wall Street Journal.*

Clouds of Beauty

1 Peter 3:3-4
"Whose adorning let it not be that outward adorning of plaiting the hair, and of wearing gold, or of putting on of apparel; But let it be the hidden man of the heart, in that which is not corruptible, even the ornament of a meek and quiet spirit, which is in the sight of God of great price."

The painted lady butterfly gets its name from the beautiful design in orange, black and white on its wings, which average about two and a half inches in full span. Every year, millions of these little creatures are seen in full migration. The painted lady is the most widespread of all species of butterfly. It is found in Asia, Africa, and South America. The species familiar to North Americans migrates from Mexico all the way to Canada. This is a one-way trip because the female butterfly lays her eggs in Mexico and then migrates to Canada, where she eventually dies. The following year, in one of the great mysteries of migration, each newly hatched butterfly in Mexico knows exactly when and where to migrate.

Some years, there are more painted lady butterflies than others. Dennis Murphy, director of the Center for Conservation Biology at Stanford University, has pointed out that the migrating population can be a thousand times greater in an exceptional year compared to normal years. Dr. Murphy believes that these unusually large populations are the result of drought conditions that also produce a greater crop of weedy thistles. The painted lady butterflies live on these thistles and never threaten any other crop.

God's design of the painted lady butterfly bears witness to His work as an artistic Creator who paints the landscape for thousands of miles with these beautiful creatures and a logical Creator who maintains the balance of nature.

Prayer: I thank You, Lord, for the lavish beauty You have created and shared so generously with all, rich or poor. Help me to be better able to share the beauty of Your salvation with those whose lives I touch. Amen.

Ref: "Butterflies are Free, and Driving in the West is Messy." *Star Tribune*, April 24, 1992, p. 7a.

The Dynamic Rain Forest

Psalm 50:10-11
"For every beast of the forest is mine, and the cattle upon a thousand hills. I know all the fowls of the mountains: and the wild beasts of the field are mine."

The environmental movement has rightfully shown concern about the South American rain forest. While it is true that the entire rain forest system is a delicately balanced harmony of thousands of subsystems, it is not true that it has taken millions of years to evolve. Each living thing forms part of the environment for other living things; therefore, adaptation is often seen as a dynamic process seemingly taking much longer than might be imagined. This is why the rain forest is spoken of as a fragile ecosystem.

Anthropologist Anna Roosevelt of the Field Museum of Natural History in Chicago has found, in Brazil, the oldest evidence of pottery in the Americas. The site where the pottery was found supported a large and industrious population. Roosevelt points out that the descendants of these ancient peoples conquered large parts of South and Central America. They cleared rain forests, built cities, and spread their culture and industry. After the invasion of the Europeans in the 1500s, these cultures died out. and the rain forests again took over. Thus, today's rain forests are likely only a few hundred years old, meaning that it has not taken millions of years of adaptation.

While we are not advocating greedy exploitation of the rain forest, it would seem from previous history that the most fragile element of the ecosystem is not the plants and animals but the indigenous human populations.

Prayer: I thank You, dear Father, that You have so wisely made the creation able to cope with a range of changing conditions. Help our knowledge of Your work of creation grow so that we may not abuse Your creation by either misuse or under-use. In Jesus' Name. Amen.

Ref: Thomas H. Maugn II. 1991. "Rain Forests not as Fragile as Believed, Ancient Pottery finds Hints." *Los Angeles Times,* Dec. 16.

The Giant Helper

2 Corinthians 9:10-11
"Now he that ministereth seed to the sower both minister bread for your food, and multiply your seed sown, and increase the fruits of your righteousness;) Being enriched in every thing to all bountifulness, which causes through us thanksgiving to God."

Nearly everyone has heard of the giant fungus in Michigan's Iron County, near the Wisconsin border. The 38-acre fungus weighs as much as a blue whale. Scientists estimate that it is about 1,500 years old, based on its current rate of growth.

Strange life forms such as this fungus bring out the whimsy in some people. A California health food store suggested that eating some of the fungus might help people live longer. Since we at Creation Moments can't resist living things that can be regarded as "strange," the fungus is a natural for us. We thought that you might like to know, however, more about the *Armillarias* fungus and its unique life.

Like most fungi, *Armillarias* is made up of tiny tendrils growing almost invisibly underground. The honey mushrooms that sprout above ground are merely the fruiting bodies of the organism. *Armillarias* are commonly found in hardwood forests throughout North America. The species is territorial, which means that no two individuals will share the same area. The fungus serves a crucial purpose in the forest ecology. It breaks down dead wood for reuse by other plants. In the process, it makes carbon dioxide for the forest plants to turn into oxygen. Without the *Armillarias*, the forest would eventually die, choked with dead wood. God has designed the *Armillarias* to be a primary recycler of waste wood in the forest.

Prayer: *Father in heaven, You know our needs even before we do. I thank You that You are so generous to us. Forgive me for complaining and for having a thankless spirit. When my need seems gigantic, help me to remember that You love me and that no need is so large that You cannot provide more than enough to satisfy. In Jesus' Name. Amen.*

Ref: John Flesher. 1992. "Humungous Fungus Tickling Funny Bones." *Kalamazoo Gazette*, April 3, p. A5.

Human Magnets

Romans 7:6
"But now we are delivered from the law, that being dead wherein we were held; that we should serve in newness of spirit, and not in the oldness of the letter."

Do you know that you have something in your head besides bone, soft tissue and blood? You have rocks in your head! Well, sort of.

Scientists recently announced that they have discovered that the human brain is laced with tiny magnetic particles made of magnetite. Magnetite is the same mineral that occurs naturally in lodestones. While the particles are distributed throughout the brain, they seem to be more concentrated in the membrane that encloses the brain. Scientists said that they had not seen the particles before because they are so tiny. The smallest are about a millionth of an inch in diameter while the largest are a hundred-thousandth of an inch in diameter. Added together, all the particles in the average brain amount to only one millionth of an ounce.

The particles are identical to those found in many bacteria, pigeons, salmon and whales. The particles help those creatures navigate, using the Earth's magnetic field. Scientists are not prepared to say that they serve the same purpose in humans. Instead, they continue to study the particles to learn what purpose they serve.

At the time of the creation, the Earth's magnetic field was several times stronger than it is today. When Christ walked this Earth, the magnetic field was twice as strong as it is today. Perhaps in that stronger magnetic field, human beings were easily able to keep track of their direction, something few people can do today.

Prayer: I thank You, Lord, for all of the talents and abilities You have given me. Forgive me for those times when I have thought that I had few talents or abilities with which to serve You. Help me to make better use of Your gifts in Your service. Amen.

Ref: Thomas H. Maugn II. 1992. "Homing Device Found in Human Brains." *The Beacon Journal*, May 12, p. A5.

Ancient Eye Surgery

1 Chronicles 16:29
"Give unto the LORD the glory due unto his name: bring an offering, and come before him: worship the LORD in the beauty of holiness."

Man has always treasured his ability to see. And since man was created as a highly intelligent creature rather than a simple primitive, it seems that man has practiced eye surgery as far back as we have records.

Eye surgery is no more a modern development than malpractice penalties. Ancient Egyptian eye doctors applied various kinds of salves to the eye. Some of their stranger salves contained mud from the Nile River and crocodile dung. Before you laugh, you should know that modern eye specialists admit that these strange ingredients may have contained effective antibodies to relieve infections. Eye surgeons were busy in India six or eight centuries before the birth of Christ. A surgeon named Susruta performed a surgery in which he pushed an opaque lens that sits behind the pupil of the eye into the interior of the eyeball; there, the lens would no longer block vision.

The oldest record of eye surgery, however, dates to 4,000 years ago. The Code of Hammurabi said that the surgeon who saved a man's eye was to be paid the same fee that was given to a doctor who saved a man's life. On the other hand, if the surgeon caused the eye to be lost, the law required the surgeon's fingers to be cut off. There were lesser penalties if the patient was merely a slave.

History records that man has always been an intelligent creature with a surprising amount of knowledge. Although sometimes some has been lost, knowledge is a gift from God.

Prayer: Dear heavenly Father, I thank You for a world in which there is beauty to touch our senses and give us delight. I ask that You would make me more appreciative of the beauty to be found in the forgiveness of my sins because of Christ's sacrifice and in serving You in holiness. In Jesus' Name. Amen.

Ref: "Looking at Way Back Then." *Science News*, Nov. 24. 1984, p. 329.

How Old Are Fossils?

1 Timothy 6:20
"O Timothy, keep that which was committed to thy trust, avoiding profane and vain babblings, and oppositions of science falsely so called:"

How long does it take to make a fossil? Would you believe that when the correct natural conditions are duplicated in the laboratory, the process only takes a few days to get underway?

Anything that was once alive can become a fossil. Bone, hair, feathers, and even cloth can fossilize. The conditions must be just right, however, or the candidate simply decays. This is what happens to most things that were once alive. To become fossilized, the candidate for fossilization has to be protected from air and other things that could cause decay. Then the original once-living molecules are replaced by molecules of silica before they can decay.

We are often given the idea that it takes millions of years for something to be fossilized. Scientists who believe in creation have never accepted that fact because they do not believe that the world is that old. Now new research into how things fossilize shows that believers in a young creation were right.

Scientists made their discovery while they were studying fossilized silk fabric from a human burial site. They discovered that if copper was buried with the person, the copper atoms would be washed into the silk by water and begin to deposit within the fiber. The result was a fossil that even showed the detail of the original threads in the cloth. In lab tests, researchers found that it only takes a few days for a considerable amount of copper to deposit in the silk.

The truth is, you could be older than a fossil!

***Prayer:** Dear Father, I thank You for the work of scientists who are showing that true science does not contradict the truth of Your Word. Prosper their work and give them the support of Your people so that we can make a better witness to those who are hindered by stumbling blocks raised up by false science. In Jesus' Name. Amen.*

Ref: Stefi Weisburd. "Fossils of Fabrics and Fibers." *Science News*, Vol. 126, p. 328.

Seismic Frogs

Psalm 65:6-7
"Which by his strength settest fast the mountains; being girded with power: Which stilleth the noise of the seas, the noise of their waves, and the tumult of the people."

The male white-lipped frog of Puerto Rico communicates using more than a variety of sounds. The frog also sends seismic signals through the ground. Researchers studying the frog have learned that it is more sensitive to seismic signals than any other creature.

The male white-lipped frog communicates using chirps, chuckles and thumps on the ground. The male begins to establish his territory by sending out a rapid stream of chirps. When he is sitting on the mud, each chirp begins with a thud. The thud is a thump on the ground that is strong enough to travel for some distance. It's thought that he does this by inflating his throat pouch so rapidly that it thumps on the ground, setting up a seismic wave.

The thump is an important part of the frog's territorial strategy. While his chirps tell other males his direction, they don't let the others know how far away he is. However, the seismic wave travels through the ground at a different speed than the chirp travels through the air. This difference allows the other males to know both the distance and direction of the male by the different times at which they hear the thump and the chirp. The other males respond to the chirps and thumps with longer sounds called chuckles.

Who taught the frog this complicated, precise system? Who taught him how to measure distance by using both airborne and seismic waves? Only our wise Creator could have done this.

Prayer: I thank You, Lord, for the many surprising ways in which Your creation shows forth Your wisdom and power. Help me to more clearly see and appreciate the wonderful things with which You have filled the creation. Amen.

Ref: J. A. Miller. 1985. "Frog Talk: Chirp, Chuckle and Thump." *Science News*, Vol. 127, Jan. 12, p. 21.

Keeping Plants in the Dark

Psalm 150:2
"Praise him for His mighty acts: praise him according to his excellent greatness."

All green plants need light to make them grow. Different plants need differing amounts of light. However, scientists have discovered a plant that is a champion at growing in dark conditions. That plant is red algae. It was discovered growing 268 meters below the surface of the ocean. So little light penetrates that deep that it would look absolutely dark to us. Scientists estimate that the light intensity at that depth is 0.0005 as much as at the surface!

How does this plant live in conditions that would quickly kill any other plant? They are designed to be 100 times more efficient at catching and using light than shallow water plants. Part of this ability comes from the unique structure of the plants. Plants typically line their cell walls with calcium, which prevents some light from getting into the cell. These plants only line the vertical walls of their cells with calcium, so there is little to prevent light from entering the cells of the plant. Additionally, the plant's cells are stacked so that the light can penetrate deep into the plant.

Good engineering, not chance, created this plant. Evolution can offer no reason for this plant to develop. But the Creator's unlimited imagination and abilities easily explain why this plant exists.

Prayer: *I thank You, dear Father, for the excellence of all of Your workmanship. I ask for the patience and wisdom to put more excellence into everything that I do so that my work and life may better glorify You. In Jesus' Name. Amen.*

Ref: S. Weisburd. "The World's Deepest-dwelling Plant." *Science News*, Vol. 127, p. 4.

Design, Not Luck

1 Timothy 4:7
"But refuse profane and old wives' fables, and exercise thyself rather unto godliness."

The clover family of plants is a large family with members all over the world. Members of this family have many forms, from the white and purple clovers that grow in the lawn to the impressive acacia and mimosa trees. However, most types of clovers live in North America.

White and sweet clovers are particularly popular with honeybees and other pollinating insects. The sugar content of the nectar produced by these clovers is 40 percent – about four times higher than popular soft drinks. Clover is also well known for its ability to restore the soil. Symbiotic bacteria in root nodules are able to fix nitrogen, which is abundant in the atmosphere, into the soil.

Ever since the days of Tom Sawyer, and probably before, young boys have looked for four-leaf clovers. Superstition says it is good luck to find a four-leaf clover. Since God is completely in charge of every detail of the creation, there is, of course, no such thing as luck. Four-leaf clovers do present an interesting variation on the usual three-leaf clover. They are fairly simple to find if you know the trick. The principle behind that trick is not luck, but genetics. You are most likely to find a four-leaf clover in a patch of clover where the leaves are not evenly shaped in the normal way. When you find a four-leaf clover in the patch, keep looking – there are probably more.

Clover has been designed by the Creator to serve man in many ways. Luck had nothing to do with it.

Prayer: I thank You, Lord, that You have not abandoned us to be the victims of luck and blind chance. Forgive me for the times that I have thought as the world does about luck, and help me to see more clearly Your involvement in my life. Amen.

Ref: Steven D. Garber. 1987. *The Urban Naturalist,* pp. 22-25.

The Deep Sea Specialist

Psalm 77:19
"Thy way is in the sea, and thy path in the great waters, and thy footsteps are not known."

The leatherback is the most massive living reptile. The largest individual ever recorded weighed over 2,000 pounds. Unlike other sea turtles, the leatherback spends nearly its entire life in deep water, coming ashore only to lay eggs. After being hatched on shore, the leatherback heads for the sea. If leatherbacks are like other sea turtles, the hatchlings won't return to land for 20 to 50 years to lay their first eggs. At egg-laying time, the female comes ashore at night, digs a nest in the sand and lays her eggs. When she is finished, she covers them and returns to the water. Ten days later, she will return and lay another nest of eggs. She may do this up to eleven times during one season. The average female will lay 120 pounds of eggs in one season.

The leatherback's favorite food is jellyfish. The eating equipment of the leatherback is perfectly designed to eat jellyfish. The sharp, cusp-like fangs on its jaw interlock in just the right way to enable it to eat bites out of jellyfish that are as large as three feet across. Its digestive tract lacks a well-defined stomach. Digestion is best done in the small intestine because jellyfish are mostly water.

The leatherback turtle has been referred to by scientists as an ocean specialist. That praises the Creator, whether or not those researchers realize it.

Prayer: *I thank You, Lord, that Your handiwork in creation is so amazing that even unbelievers praise it. Help me to witness more effectively so that I may be Your instrument to bring people the complete message that You have also died so that we can be forgiven, and You have been raised so that we can live new lives. Amen.*

Ref: Scot A. Eckert. 1992. "Bound for Deep Water." *Natural History*, March, pp. 29-35.

The First Xerographer

2 Chronicles 26:15
"And he made in Jerusalem engines, invented by cunning men, to be on the towers and upon the bulwarks, to shoot arrows and great stones withal. And his name spread far abroad; for he was marvellously helped, till he was strong."

Xerography is the process used in most copying machines today. The page you want copied is exposed to the bright light in the machine. That light enables the machine to make an image of your original, using electrostatic charges to record the light and dark spots on your original. The toner – that fine black powder – is then drawn to those spots on the copier paper in exactly the same spots that were dark on your original. After the powder is fixed in place, your copy – perfect in every detail – pops out.

Most of us consider xerography to be a modern work-saving wonder. However, xerography may not be such a modern invention. Researchers at Stanford Research Institute now believe that xerography was invented and first used almost 150 years ago!

That's when Louis Daguerre introduced the photographic process named after him. The earliest photographs we have are daguerreotypes. These sharp, crisp pictures used silver and iodine vapor to fix an image on a plate. The process used the xerographic principle of setting up electrostatic charges on a plate wherever light struck it. It was the sharpness in the detail of the daguerreotypes that convinced researchers that Daguerre's process was xerographic.

Unfortunately, his process was cumbersome and dangerous. Nor did Daguerre understand how the process worked. If he had, Abraham Lincoln might have had a copier in the White House!

That humans are ever inventive and creative arises from the fact that they are the creation of an unlimited and wonderful Creator.

Prayer: *I thank You, Lord, for the gift of inventiveness that You have given to man. I ask that we may always be guided to use our creativity and inventiveness to Your glory. Amen.*

Ref: Bower, Bruce. 1985. Picturing an electric look. *Science News*, v. 127, Feb. 2, p. 74.

The Flowering Chameleon

Matthew 6:28-29
"And why take ye thought for raiment? Consider the lilies of the field, how they grow; they toil not, neither do they spin: And yet I say unto you, That even Solomon in all his glory was not arrayed like one of these."

What color is the scarlet gilia? Well, that depends on where and when it's blooming. Despite its name, the scarlet gilia can be red, pink, and even white.

Near sea level, the flowers remain red all season. This is because the flower is pollinated throughout the summer by hummingbirds who are drawn to the red flowers. At higher elevations, the hummingbirds leave as summer moves into August. As they leave, the plants change their red flowers to pink and, later, white flowers.

The higher-elevation plants change over to white flowers because they have a new pollinator. As the hummingbird population decreases, hawkmoths take over. During this transition, the flower produces pink blossoms as it invites both of its pollinators. Once the hummingbirds are gone, the hawkmoth takes over. The hawkmoths pollinate at night, so they prefer white flowers, which are easier to see in the dark. Scientists were amazed by the fact that the color changes take place exactly when pollination shifts from hummingbird to moth. As one startled scientist put it, these plants cannot be considered passive in their environment.

Who taught the scarlet gilia when and how to change the color of its flowers? Who taught it that hummingbirds will be attracted to red flowers? Or that hawkmoths will be attracted to white flowers? The scarlet gilia may be clever, but only our all-wise Creator could have given this knowledge and ability to the plant.

Prayer: Dear Father, I thank You that the wisdom with which You designed the creation is there for all to see. As I strive against the world's denial of Your intimate involvement in Your creation, help me to see more clearly that the creation is Yours and that it glorifies You. In Jesus' Name. Amen.

Ref: Bennett, D. D. Scarlet gilia: Flowering Chameleon. *Science News*, Feb. 1985, p. 69.

Learning from Experience

Proverbs 22:6
"Train up a child in the way he should go: and when he is old, he will not depart from it."

Christians who believe in the biblical account of creation are usually strongly pro-life. This is because they realize that the Creator uses a woman as His workshop to form another human being to love.

Polls repeatedly show that only a minority of the U.S. population favors free and unrestricted abortion for all purposes including birth control. Pro-abortionists have hoped to change this by training the next generation to accept their position. That's why they were quite upset when they saw the results of their latest study.

In 1969, 75 percent of all high school freshmen considered themselves "pro-choice." However, in 1983, only 37 percent considered themselves "pro-choice." In a more recent study, pro-abortionist educators discovered that teenagers associate abortion with terms like "killing," "death," and "pain." Teenagers also commonly consider abortion to be harmful to their own well-being. Pro-abortionists blame pro-lifers for these perceptions of abortion among young people. Wanda Franz, president of National Right to Life, notes that pro-lifers can't take all the credit. She points out that young people have abortions and then talk to each other about their experience. They are giving each other first-hand information about the cause of the death of one-third of their generation.

Those of us who believe in life should not consider this a victory. Rather, it is a signal that teenagers are an important and winnable front in the war against death.

Prayer: Lord of life, I ask that You would encourage and strengthen those who are working to protect life. Especially strengthen and encourage those young people who are encouraging their friends to recognize the sanctity of human life. Amen.

Ref: Franz, Wanda. Pro-Abortionists Worry about the Pro-Life Attitudes of Teenagers. *National Right to Life News*, June 2, 1992, p. 3.

Smart Lobsters

Proverbs 21:24
"Proud and haughty scorner is his name, who dealeth in proud wrath."

Do spiny lobsters know that they are a delicacy to many, if not most, of the carnivores on earth? New research into the social habits of spiny lobsters shows that much of their interaction with each other is based on the fact that they have learned to cooperate in order to survive their many predators.

Logic told researchers that a lobster would have a better chance of escaping predators by hiding in a small hole all by itself. Yet, lobsters are often found in groups of ten or more. Researchers wanted to know if the lobsters knew something they didn't.

They found that a lobster that was all alone would indeed hide in the smallest and most secure enclosure it could find. When more lobsters were present, however, they preferred to gather in larger enclosures. Lobsters that were not under threat from a predator would seek out roomier lairs. When a threat was present, they would still congregate, only in the smallest possible hiding place. This still made them more accessible to predators than if each was alone in a small hole. Researchers soon discovered that this disadvantage was outweighed because the larger group could more easily spot danger and alert the others. When threatened, the lobsters would face their attacker and try to fend it off by swatting it with their spiny antennae. The lobsters' combined strategy was more effective than the strategy that seemed logical to the scientists.

This example shows us the limitations of science. The Creator made lobsters even smarter than scientists when it comes to lobster self-protection.

Prayer: Dear Father, there is much pressure today to accept what science says, even when it disagrees with Your Word. Help me to see where and how this pressure affects me so that I may identify the problem and have a stronger, more informed faith. In Jesus' Name. Amen.

Ref: Ezzell, Carol. 1992. "Spiny lobsters: there's safety in numbers." *Science News*, v. 141, May 30, p. 357.

Sepphoris

Matthew 6:2
"Therefore when thou doest thine alms, do not sound a trumpet before thee, as the hypocrites do in the synagogues and in the streets, that they may have glory of men. Verily I say unto you, They have their reward."

Nazareth, the town in which Jesus grew up, is often portrayed as a tiny backwater town. As a result, Jesus is often portrayed as an unsophisticated man. He has been painted many times as someone who could more easily relate to farmers and shepherds than to the city dweller. Biblical researchers point out that this is not an accurate picture of Jesus of Nazareth.

It is true that Nazareth was only a small village of 400 people in Jesus' time. However, only four miles away stood Sepphoris, the main city and capital of Galilee and Perea. Today, Sepphoris is nothing but ruins. But in Jesus' day, it was a bustling city of 30,000, fully linked to the trade routes of the empire. It was the royal residence of Herod Antipas. It also boasted a 4,000-seat theater, public baths, archives, gymnasium, water works, and other public buildings. In other words, rather than a tiny backwater town, Nazareth was a suburb of a major Roman capital!

Researchers have concluded that Jesus lived in a more urban and sophisticated culture than many have believed. This explains why many of Jesus' parables talk about the policies of kings, tax collectors, wealthy landlords, and even actors. Yes, actors. In His teaching, Jesus often referred to hypocrites. The word "hypocrite" comes from the Greek word for stage actor. Stage actors, of course, pretend a role or play a part.

Not all scientists challenge Scripture. In this case, archaeology is helping us better understand the context of our Lord's earthly life.

Prayer: I thank You, Lord, that there are scientists who are working to help us better understand the Bible rather than to discredit it. Bless and prosper such work and those who do it so that Your Name may be more greatly glorified among us. Amen.

Ref: *Biblical Archaeology Review*, May/June 1992, p. 52.

Stone Makers

Ephesians 2:8-9
"For by grace are ye saved through faith; and that not of yourselves: it is the gift of God: Not of works, lest any man should boast."

Modern sophisticated laboratories have produced many specialized materials with amazing qualities. Yet, many creatures make ceramic materials that are stronger and tougher. If a crack starts in a piece of our ceramic, over time it grows. The same crack will not grow in ceramic made by the oyster.

Human and animal bodies custom-form teeth for a variety of purposes. When science makes such materials, it uses heat, chemicals and high pressure. One scientist noted that it would take experts in five or six scientific disciplines to figure out how living things form these materials.

Scientists have been trying to learn how creatures that make stony materials do it. Most of what they have learned has left them astonished at the quality of the materials, even if they have learned little about the process. They believe that special cells in the body use proteins and other large molecules to form non-organic molecules into stony materials. It appears that many creatures can control the formation of crystals. For example, in the abalone shell, crystals and glue are stacked in brick-and-mortar fashion in such a way that they become stronger under pressure. At the same time, the structure allows them to be formed into intricate shapes.

Our best human efforts cannot match God's gifts. Our best human behavior cannot make us good enough for God. Only the forgiveness of sins that comes by grace through faith in Christ can do that.

Prayer: Our dear heavenly Father, I thank You that You have sent Your Son, Jesus Christ, to work out my salvation. I thank You for Your forgiving grace that assures me of the forgiveness of my sins and Your love. In Jesus' Name. Amen.

Ref: Pennisi, Elizabeth. 1992. "Natureworks: making minerals the biological way." *Science News*, v. 141, May 16, p. 328.

Hairy Drug Factories

Matthew 13:22
"He also that received seed among the thorns is he that heareth the word; and the care of this world, and the deceitfulness of riches, choke the word, and he becometh unfruitful."

We are all familiar with the fact that the very tips of most plant roots have tiny hairs. These hairs seek out water and nourishment from the soil and absorb them for distribution in the plant. Few people beyond some biologists and chemists are aware that these delicate portions of some plants are able to produce a large variety of important medicines. Often, the hairy parts of plant roots are able to make medicines that even our best chemists and genetic engineers cannot make.

The hairy roots of plants have been used to make certain medications for over two generations. But it was only recently that scientists discovered that by infecting seedlings with a certain bacterium, they could produce versions of plants that were nothing but hairy roots. Another benefit of this new procedure was that the hairy roots never stopped growing or making medicine.

Periwinkle hair roots make 75 different alkaloids, including caffeine, nicotine, and two medications important in cancer treatment. Antidotes for motion sickness and nerve gas poisoning are made by henbane roots. Gourd roots make important agricultural chemicals. Relatives of the Chinese cucumber make anti-fungus chemicals. And, as many gardeners know, marigold roots make chemicals that act against nematodes and fungus.

While human sin has made the creation less than perfect, our merciful Creator has, with foresight, provided solutions to many of our earthly problems.

Prayer: Heavenly Father, I thank You that You have so generously provided us with solutions to so many of the problems we face because sin has ruined Your creation. Never let my earthly problems, or their solutions, distract me from my need for forgiveness of my sins through Jesus Christ. Amen.

Ref: Pennisi, Elizabeth. 1992. "Hairy harvest: bacteria turn roots into chemical factories." *Science News*, Vol. 141, May 30, p. 366.

Natural Musical Abilities

Ephesians 5:19-20
"Speaking to yourselves in psalms and hymns and spiritual songs, singing and making melody in your heart to the Lord; Giving thanks always for all things unto God and the Father in the name of our Lord Jesus Christ;"

We have previously reported on research that shows that human beings are born with a natural language ability. We are not only born with a hunger to learn language, we are also born with certain expectations about what language is. This fact can only serve as a witness to the Creator and His genius.

Now research into how infants respond to music has shown that we are also born with a readiness to learn the basics of what music should sound like. Western major and minor scales follow a certain mathematical pattern. However, the Javanese pelog scale is based on much more complex mathematical relationships. Those who have grown up with Western musical patterns find that music written in the pelog system sounds weird to their ears. When does this sense that tells us that certain mathematical patterns of notes don't fit together develop?

To find out, researchers tested adults and six-month-olds. Adults can easily identify notes that don't seem to fit in a string. The infants had been trained to look at a speaker when a note was not in tune. When they were right, they were rewarded by watching the actions of an animated toy. Infants that had been trained in both Western and pelog scales did better at identifying mistuned pelog notes than untrained adults.

Music is a gift of God. We find music used and referred to from early in Genesis to the end of the book of Revelation. The greatest use we can make of music is in praise of our Creator.

Prayer: I thank You, Lord, for the gift of music. It makes our hearts glad and gives us yet another way to worship You. Help Your people always make use of music. Amen.

Ref: "Infants tune in to the sounds of music." *Science News*, Vol. 138, p. 46.

Glow-in-the-Dark Flowers

Psalm 96:12-13a
"Let the field be joyful, and all that is therein: then shall all the trees of the wood rejoice Before the LORD:"

Researchers at Stanford University surprised scientific colleagues several years ago when they proved that plants were not ... well, vegetative. We think of plants as things that sit and grow, not as living things that can react to their environment. Now researchers at the University of Edinburgh, Scotland, have shown that plants not only react to the environment, they react with the same immediacy as animals.

The Edinburgh researchers knew that plants react to a breeze or other wind movement by adding calcium to their cell walls. Plant cell walls typically have calcium in them. The calcium acts as an internal skeleton, giving strength to the plant. When stressed by wind currents, plants strengthen themselves by adding more calcium to their cell walls.

Researchers used a novel method to study how quickly plants react. They added to the plants jellyfish genes that bind to calcium and glow blue as the calcium level increases. Then they squirted the plants with puffs of air. The increased blue glow showed that plants react almost immediately to air movement by adding calcium to their cells. The researchers have been contacted by private companies that want to find out whether this technique can be used to develop glow-in-the-dark flowers or glowing grass to plant around airport runways.

When God created plants, He gave them abilities that seem surprising to those who think that they are a simple form of life.

Prayer: I thank You, Lord, that You have filled Your creation with wonders beyond counting. I ask that You would help me identify the effects of evolutionary thinking that devalue the wonder of Your creation in my thinking so that I may better praise You. Amen.

Ref: "Glow little stressed plant, glow." *Science News*, Vol. 141, June 6, p. 379.

Taking a Bite Out of Stress

Psalm 130:4
"But there is forgiveness with thee, that thou mayest be feared."

Research has repeatedly shown that dogs can be an effective sedative. Continued research has now shown that dogs can relieve more stress and promote better health than even the presence of a close friend will do.

A 1980 study showed pet owners who had been hospitalized for heart disease had a better survival rate and were healthier than heart patients without pets. Another study showed that elderly pet owners had fewer visits to their doctors than those with no pets.

New research shows that dogs lower blood pressure and other bodily responses to stress even better than the presence of a good friend. Researchers subjected volunteers to stress by asking them to count backwards by threes, then sevens, then thirteens and seventeens. As they counted, their stress levels were measured. The tests were done in volunteers' homes. Some volunteers had dogs, others had a friend sitting nearby. Those with friends nearby did their mental arithmetic more rapidly, but less accurately, than those with dogs. Those who had earlier performed well in the presence of their dog did more poorly when the dog was removed and a friend was nearby.

Researchers explained these findings by pointing out that dogs provide unconditional support. God offers us the unconditional forgiveness of our sins, by grace through faith in His Son, Jesus Christ. That's why Scripture connects the forgiveness of our sins with peace in this life and forever.

Prayer: *I thank You, my dear heavenly Father, for Your unconditional forgiveness of my sins. If it had been up to me to earn Your forgiveness, I would be lost forever. Grant me a greater faith and a more full realization of Your peace. In Jesus' Name. Amen.*

Ref: Bower, Bruce. 1991. "Stress goes to the dogs." *Science News*, Vol. 140. Nov. 2, p. 285.

The Mouse with Radiation Protection

1 Peter 5:6-7
"Humble yourselves therefore under the mighty hand of God, that he may exalt you in due time: Casting all your care on him; for he careth for you."

Most mice are nocturnal. The African four-striped grass mouse, however, operates in the fierce equatorial sun.

The African four-striped grass mouse goes out in the hottest sun to eat roots, green grass and seeds. It may do its shopping on the ground or climb a low branch to select food from a low shrub. The chipmunk-like mouse is found throughout eastern and southern Africa, where the sun is the strongest. Near the equator, the powerful sunlight also includes powerful doses of ultraviolet radiation.

The African four-striped grass mouse is more equipped to handle this harsh environment than most other creatures. Beneath its fur, the mouse has melanin-pigmented skin. Melanin is the pigment that turns our skin brown when it's exposed to the sun. Melanin protects the skin from sun damage, including ultraviolet radiation. The grass mouse also has an additional melanin-pigmented layer between its skull and its outer skin. This layer provides additional protection for the mouse's brain. Only three other African rodents have this special protective helmet. White tent-making bats of South America have similar protection. While these bats sleep during the day, they sleep in a curled position that exposes their heads to the South American sun.

This special protection was obviously designed keeping in mind the danger of over-exposure to ultraviolet radiation. Humans have only recently learned about these dangers. We can safely guess that the animal world doesn't know about it at all. But our all-wise Creator does.

Prayer: Lord, I thank You that You have commanded us to pray and assured us that the Father hears us and knows our need, even before we ask. I thank You that I am not the product of a mindless, uncaring universe. I thank You for Your love. Amen.

Ref: Carter, Laura S. 1992. "Ban the Soleil." *Natural History*, June, p. 76.

Reptilian Housing Development Expert

Psalm 40:5
"Many, O LORD my God, are thy wonderful works which thou hast done, and thy thoughts which are to us-ward: they cannot be reckoned up in order unto thee: if I would declare and speak of them, they are more than can be numbered."

He looks slow and unimportant. He spends 90 percent of his time underground. But the gopher tortoise constructs a housing development that is home to over 360 species. In addition, his tunnels provide emergency shelter to many other creatures during forest fires.

Gopher tortoise burrows are often 30 feet long and may extend to a depth of 15 feet. The record length of a tunnel is nearly 50 feet. One tortoise may dig or enlarge many tunnels over its 60-year lifetime. Obviously, the tortoise is well-designed for digging, with spade-like front legs. The tortoise is a vegetarian. It is found in temperate climates, so its burrow serves to protect it from winter temperatures that are dangerous to reptiles. In addition, the tortoise has a large urinary bladder touching the back of its shell that acts as a hot-water bottle. The tortoise heats the water by aiming its back at the sun.

Because the tortoise can weigh over 10 pounds, its tunnels are large enough to accommodate many creatures. The Florida mouse will dig several tunnels off the main tunnel that it calls home. Insects, lizards, snakes, alligators, bobcats, and even wild turkeys, burrowing owls, wrens and robins use the burrow for shelter or to look for food. Abandoned burrows may be used by foxes, skunks and other mammals for dens.

The gopher tortoise, now endangered, has been designed by the Creator to make the forest community richer and more diverse.

Prayer: Father in heaven, I ask that you would move those who can do something for the gopher tortoise to take reasonable steps to preserve this wonderful creature in forested areas. In Jesus' Name. Amen.

Ref: Burke, Russell L. 1992. "Multiple Occupancy." *Natural History*, June, p. 9.

Music Grass

Genesis 4:21
"And his brother's name was Jubal: he was the father of all such as handle the harp and organ."

Reed instruments like the oboe and clarinet trace their ancestry back to instruments made by the Egyptians more than 4,000 years ago. An Egyptian relief dated to about 2700 B.C. shows a clarinet-type instrument.

This means that humans have been dealing with the problem of making reeds work in reed instruments for thousands of years. Natural reeds can crack, break and make awful honking noises as they age and wear out. Today, many musicians use plastic reeds. Still, professional musicians reject the sound of plastic reeds as inferior for public performance. As one professional musician put it, "They sound kind of like a duck with laryngitis."

Reeds are fashioned from the giant reed plant. This large grass grows to a height of seven or eight feet. Reeds are made from the sections between the nodes. However, a seven-foot stalk may have only 25 sections that are the right dimensions for an oboe reed. After harvesting and curing, a process that takes nine months, sections are selected on the basis of size and color. One plant may produce only one oboe reed. The goal is to obtain a reed that has a crisp tone over a wide range of notes. Out of a box of 100 reeds, which can cost $150, the professional musician may find two that he considers excellent.

Making music is one of the earliest human activities described in the Bible. The mention of music early in Genesis shows us that the earliest people were just as human as people today.

Prayer: I thank You, dear heavenly Father, for the gift of music. Help us to use music to praise You and rejoice in Your generous goodness to us. In Jesus' Name. Amen.

Ref: Schmidt, Karen F. 1991. "Good vibrations: musician-scientists probe the woodwind reed." *Science News*, Vol. 140, Dec. 14, p. 392.

A Not-So-Hard Saying of Jesus

Matthew 8:22
"But Jesus said unto him, Follow me; and let the dead bury their dead."

Twice in the Gospels, Jesus is asked by a disciple for a leave of absence to bury his father. In each case, Jesus tells the disciple to let the dead bury their own dead.

This has been described as one of Jesus' hard sayings, since it suggests that a son should follow Jesus rather than honor his father. The passage is usually explained as meaning that the spiritually dead – that is, those not following Jesus – should bury the physically dead. This explanation still doesn't explain the apparent advice to break the commandment about honoring our parents.

A better explanation is offered in an article in *Archaeology and Biblical Research*. Archaeologist Gordon Franz explains Jewish burial practices at the time of Jesus. At death, the body would be placed in the family burial cave. After about a year, the body would have decomposed. Then the final act of mourning would take place. The bones were placed in a chest in what was called a secondary burial. According to the rabbis, the decomposition of the flesh was a person's final atonement for sin. Once this atonement was made, the bones could finally be laid to rest.

Franz says it is this secondary burial for which the disciples wanted leave time. Jesus' reference to the dead burying the dead was a reference to the other corpses in the burial cave. He was teaching that the rabbis' teaching about atonement was wrong. Atonement is found only in the saving work of Jesus Christ – the very message that the disciples should have been proclaiming.

Prayer: I thank You, Lord Jesus, that in Your innocent suffering and death You made full and complete atonement for my sin. Equip me to better tell others that You have made our peace with God, and let my life reflect Your love. Amen.

Ref: Franz, Gordon. 1992. "Let the dead bury their own dead." *Archaeology and Biblical Research*, Vol. 5, n. 2, Spring, p. 55.

Tiny Superlatives and God's Love

Psalm 10:14
"Thou hast seen it; for thou beholdest mischief and spite, to requite it with thy hand: the poor committeth himself unto thee; thou art the helper of the fatherless."

Hummingbirds are among the most exquisite jewels in God's creation. Many of their activities and habits seem almost unreal.

I know of no one who has ever tired of watching a hummingbird hover or fly backward. Our sense of wonder is not decreased because we understand how the bird can do these tricks. Perhaps an even greater wonder is how these tiny, fragile creatures can make their way in a large and often hostile world.

Consider the broad-tailed hummingbird, for instance. This tiny hummingbird migrates 1,200 miles between Mexico and the Colorado Rockies each year. Its nesting site in the Rockies is filled with predators. In this setting, the mother searches for a protected branch where she will build her fragile nest. It's best if there is another branch right above the nesting branch to provide shelter from the elements as well as visual cover from hawks and blue jays. She weaves her nest, about the size of half a golf ball, from spider webs and down from plants. It will take two and one-half weeks for her two pea-sized eggs to hatch. In the meantime, the tiny mother, who weighs less than one-eighth of an ounce, must keep herself and her eggs warm in temperatures that can drop below freezing.

God has equipped the delicate hummingbird with intelligence and excellent flying ability, so that even though it is tiny and weak, it can make its living. The continued existence of this fragile creature glorifies its Creator, not the principle of survival of the fittest!

Prayer: I am filled with wonder, Father, at how You have designed and cared for the hummingbird. When I am tempted to act by my own strength instead of Yours, remind me of Your care for the fragile hummingbird. In Jesus' Name. Amen.

Ref: Calder, William A. 1992. "Ten years on an aspen branch." *Natural History*, July, p. 4.

Masters of Mimicry

Isaiah 40:28
"Hast thou not known? hast thou not heard, that the everlasting God, the LORD, the creator of the ends of the earth, fainteth not, neither is weary? there is no searching of his understanding."

Though they are only insects, stick insects seem to have an amazing knowledge of their surroundings. They are geniuses at using their predators' weakness for their own protection.

Stick insects are leaf eaters. Many of them look like little twigs. The longest insect in the world is a walking stick from Borneo that reaches a length of 13 inches. A tropical stick is as thick as a finger and the same color as the bamboo on which it is often found. It also has swollen ridges that make it look even more like bamboo. Other types have leafy flaps that match the leaves of the plants they frequent. Some stick insects go even further. They lay eggs that look exactly like the seeds of the plant on which they feed. A stick that's found in New Mexico glues its eggs to grass stems. The position and shape of the long pointed eggs exactly mimic the seeds of that species of grass.

While they don't fly, some sticks have brightly colored wings that they can rapidly unfurl. They use this ability to startle away nervous birds who might want to investigate them for lunch. Sticks will often sway with the breeze to make their illusion more effective. Some sticks will sit motionless for hours. Some birds understand this and will closely examine non-moving twigs. Sticks will often refuse to move even when being carried away by a bird. This tactic has fooled many birds into thinking it really was a twig and releasing it.

Stick insects are clearly not the result of mindless chance. Their intelligence was created by the Creator of the universe.

Prayer: Dear Lord, I can never thank You enough that You are the Creator and in charge of the universe. Many in our world do not know this and live in hopelessness. Help us to reach them in their hopelessness with Your good news. Amen.

Ref: Sivinski, John. 1992. "When is a stick not a stick?" *Natural History*, June, p. 30.

Scuba Gear for Bacteria

Job 21:22
"Shall any teach God knowledge? seeing he judgeth those that are high."

Biotechnology may be a new word to you. It's a relatively new science. Biotechnology is the science of engineering custom-made life forms. For example, biotechnologists can add to a bacterium the gene that makes fireflies glow. This procedure gives the bacterium the ability to glow.

Today, many engineered bacteria have been designed. They make medicines, hormones, enzymes and even food products. As biotechnologists are learning more about how to custom-design bacteria, they are able to make bacteria that are better able to do their assigned jobs. The efficiency of bacteria in doing their jobs can be affected by how efficiently they use their food and how much oxygen they need. Getting enough oxygen to bacteria growing in vats is often a challenge.

Biotechnologists have now discovered a bacterium that does well in low-oxygen situations. It can get by with periods of lower oxygen because it makes hemoglobin, the same molecule that carries oxygen in our blood. The bacterium uses the hemoglobin much like scuba gear to help it breathe. Biotechnologists have successfully added the hemoglobin gene to bacteria to make an enzyme used in the commercial production of high-fructose corn syrup. As a result of their increased efficiency, these bacteria are now able to make twice as much enzyme.

Even so-called "simple" living things are a symphony of carefully designed abilities. Modern science is revealing that they are intelligently designed by our Creator.

Prayer: Dear Lord, I thank You that You have allowed us to learn some of the secrets of how You designed life to work so that we may improve life here on earth. I pray that You would prosper the positive use of this knowledge and discourage any possible misuse. Amen.

Ref: Fackelmann, Kathy. 1991. "'Scuba gear' for biotech bugs." *Science News*, v. 140, p. 334.

Beautiful Loyalty

1 Corinthians 7:10
"And unto the married I command, yet not I, but the Lord, Let not the wife depart from her husband:"

Most water fowl are faithful to one mate over long periods of time. The swan, in particular, champions marital fidelity.

In North America they are called whistling swans or tundra swans. Another member of the same species is usually found in Europe and Asia, where it is called Bewick's swan. Researchers report that each has a unique pattern of yellow and black on its bill, making individuals easy to identify. This allowed researchers to trace their lives over many years.

These beautiful creatures have relatively long life spans. One individual was traced for 26 years. Typically laying four eggs in nesting season, they breed for most of their lives. They are highly territorial. In a face-off for food or nesting area, the largest male is usually the winner. When there is a disagreement, the males face each other while their mates cheer them from the sidelines. Single swans enjoying a nice discovery of food are often driven off by couples. Couples are faithful for life. Researchers say that they have recorded no cases of divorce among successfully reproducing couples. Nor do they report any case of marital infidelity among swans. If a mate is lost, the survivor often takes over a year to settle down with another mate. In one case, a survivor waited for six years.

Marital fidelity is also God's instruction to humans. Today many factors are allowed to challenge marital infidelity. Often, infidelity before or after marriage is portrayed as simply normal. But if swans can manage fidelity, we humans have no excuse for infidelity.

Prayer: I thank You, dear Father, for the gift of marriage. Help Christians, both young and old, to resist the pressures around us to misuse sex. Help us to be a positive example in a society that shows a decreasing respect for marriage and that increasingly suffers from the curses brought about by infidelity. In Jesus' Name. Amen.

Ref: Scott, Dafila. 1992. "Swans semper fidelis." *Natural History*, July, p. 26.

The Salt of the Earth

Matthew 5:13
"Ye are the salt of the earth: but if the salt have lost his savour, wherewith shall it be salted? it is thenceforth good for nothing, but to be cast out, and to be trodden under foot of men."

There are many different kinds of salts that occur naturally in our environment. Yet, only one of those salts stirs our taste buds. That salt – sodium chloride – is the tastiest of all the salts and the one that all living things need. The amazing chemistry behind this mystery shows that this is no accident.

The sodium in common table salt is crucially important in keeping the body's water in balance. That makes common table salt irreplaceable for most land animals and humans. We cannot taste anything unless its molecules are able to get into the dense network of filaments that coat our taste buds. Most salts, such as aluminum or magnesium salts, offer no appealing flavor. This is because the molecules that make up these salts are too large to get through the filaments around our taste buds.

So sodium chloride, common table salt, is a carefully designed molecule. We need the sodium, but can't really taste it. The chloride, however, is small enough to get into our taste buds and provide that salty taste. Linked together, each part of the molecule serves a crucial purpose in giving salt its flavor.

There are many beliefs among humans and many types of behavior. However, there is no other way to come to our Creator and God than through the forgiveness of sins that's found in Jesus Christ. Nor is there any other way for us to be truly God's salt in the world than by making His forgiveness, Word and will part of our very beings.

Prayer: I thank You, Lord, that even the molecular designs in Your creation serve to make life possible. I pray that my life may be so thoroughly informed by Your Word that I cannot help but be true salt in witness to Your salvation. Amen.

Ref: Ezzell, Carol. 1991. "Salt's technique for tickling the taste buds." *Science News*, Vol. 140, p. 276.

Ant Empires

Romans 13:13
"Let us walk honestly, as in the day; not in rioting and drunkenness, not in chambering and wantonness, not in strife and envying."

Like an ancient empire, there is a species of ant whose entire society is built on slavery. In fact, raids to capture more slaves are about the only thing resembling work that western slavemaker ants do.

Western slavemakers are a species of red ant. Their survival depends on making slaves out of a related species of ant. Their slave-capturing raids are dramatic. After a scout finds a suitable nest to raid, an impressive raiding party is formed. As many as 2,500 individual ants swarm into an army that can be 3 feet wide and 16 feet long. When they reach the nest, the ants spray it with a chemical that causes the inhabitants to leave. Raiders then capture the pupae and return them to the slavemaker nest. The only life the captives will ever know is one of slavery. A single slavemaker colony may steal as many as 30,000 slaves a year.

A slavemaker nest may have 6,000 slaves to serve every need of 3,000 inhabitants. Scientists have found that hunger in the slavemaker nest is the primary cause of raids to increase the slave population. Well-fed colonies engage in fewer raids. If the slavemaker colony moves, the slaves must carry each slavemaker ant to the new location.

We see a connection between the slavemakers, who work only to bring what we would regard as misery to others, and their failure to produce anything positive. This illustrates the Bible's warning that unproductive lives and strife go hand in hand.

Prayer: Dear Father, I thank You that through the innocent suffering and death of Your Son, Jesus Christ, on my behalf, I am freed from slavery to sin. Help my life to be productive for You and free from the strife that results from our sin. In Jesus' Name. Amen.

Ref: Miller, Julie Ann. 1985. "Slave-making ants rob the cradle." *Science News*, Vol. 127, p. 164.

How to Engineer a Better Turtle

Genesis 1:21
"And God created great whales, and every living creature that moveth, which the waters brought forth abundantly, after their kind, and every winged fowl after his kind: and God saw that it was good."

Leatherback turtles make use of far more of the sea than most creatures. They can dive to depths of over 4,000 feet. They prefer cold, open ocean waters but lay their eggs on tropical beaches. Not only are they found all over the world, but the same *individual* may be found all over the world. In 1970, a turtle that was tagged off the South American coast also turned up just off the West African coast, almost 4,000 miles away.

Leatherbacks are uniquely designed, inside and out, to travel the world's seas. Their bodies are teardrop shaped. This gives them nearly ideal efficiency traveling through water. In addition, they have seven ridges along the length of their upper shell, further increasing the efficiency of water flow around them. Their bodies, including the shell, are covered with a very smooth, leathery skin. Studies of leatherback hatchlings show that they use 20 percent less energy to swim through the water than any other turtle hatchling.

Leatherbacks live their entire lives in the open sea, except for a trip to shore to lay eggs every few years. They almost never stop moving. Scientists who attached radio monitors to leatherbacks were astonished to find that they swim and dive day and night, never stopping to sleep. Their front flippers are more than half the length of the entire body. They are driven by powerful pectoral muscles that can make up as much as 30 percent of the turtle's weight.

The leatherback turtle is a marvelously designed creature whose very existence glorifies our Creator.

Prayer: Lord, I thank You for the excellence You put into everything that You have made. Help my every effort that it might be my best and most excellent offering to You in thanksgiving for Your goodness to me. Amen.

Ref: Eckert, Scott A. 1992. "Bound for deep water." *Natural History*, Mar., p. 29.

Coded for a Poisonous Lifestyle

Job 28:24-25
"For he looketh to the ends of the earth, and seeth under the whole heaven; To make the weight for the winds; and he weigheth the waters by measure."

The bacterium *Bacillus cereus* likes to live in soil that would kill most other bacteria. It's bothered neither by lethal concentrations of metal nor natural antibiotics.

The bacterium is commonly found in the soil. However, in soil that overlays metal deposits like gold and copper, there may be 100,000 times as many *cereus* as normal. Soil over such deposits picks up a metallic content from the mineral below. If the metallic content is high enough, only a few plants, metal-tolerant fungi, and *cereus* can live in the soil. To make matters worse, the metal-tolerant fungi make penicillin and other antibiotics in concentrations high enough to kill bacteria. But none of this fazes *cereus*.

How can *cereus* survive under conditions that would kill any other known bacteria? Researchers believe that *cereus* performs some sophisticated chemistry that detoxifies the metal and makes the penicillin harmless at the same time. For example, in the case of copper, they believe that *cereus* removes a water molecule from the penicillin molecule. The resulting gap in the penicillin molecule traps metallic molecules in the soil and makes them harmless. They suspect that *cereus* may use different sophisticated strategies to deal with different metals. As one of the scientists pointed out, each *cereus* seems to be genetically coded for its unique location.

Cereus may be only a bacterium, but its clever chemistry and unique individuality bear witness to both our Creator's supreme intelligence and His care for each of His creatures.

Prayer: I thank You, Lord, for the wonders You have made. I also thank You that You have not simply formed the earth and then left it to spin off in space on its own. I thank You that You care about all that You have made. Amen.

Ref: Weisburd, Stefi. 1985. "*Cereus* bacteria go for the gold." *Science News*, Vol. 127, Feb. 16.

Species Confusion May Kill Ape-Man

Genesis 1:26
"And God said, Let us make man in our image, after our likeness: and let them have dominion over the fish of the sea, and over the fowl of the air, and over the cattle, and over all the earth, and over every creeping thing that creepeth upon the earth."

A debate about the definition of "species" may remove *Homo erectus* from the human evolutionary tree.

Traditional evolutionists theorize that *Homo habilis* evolved into *Homo erectus*, who evolved into *Homo sapiens*. Some evolutionists now want to reclassify *Homo erectus* as a primitive form of *Homo sapiens*. Those who don't want to abolish the *Homo erectus* classification have been supporting their position by emphasizing the differences between *Homo erectus* and *Homo sapiens*. *Homo erectus* has a considerably smaller brain than *Homo sapiens*. These creatures also had larger teeth and heavier limb bones. Scientists who want to keep the classification conclude that *Homo erectus*' characteristics are quite distinct from *Homo sapiens*'. Those who want to reclassify *Homo erectus* as *Homo sapiens* want to blur the differences that we generally use to discern between humans and apes. We need to ask, if evolutionists cannot define a species, how do they know when a new one evolves?

How different is *Homo erectus*? Dubois, who discovered the first fossils that are today recognized as *Homo erectus*, finally concluded that the fossils were only the remains of a large gibbon. Dubois had promoted the fossils as those of "ape men" for decades, before reversing his position.

The Bible leaves no room for so-called ape-men. Today's human beings are not improved ape-men, nor was Jesus Christ, Who became a man to carry our sins so that we might be forgiven and restored to God.

Prayer: I thank You, dear Lord, that You became human to carry my sin on the cross. I thank You that evolution is not true and that You originally made us distinct from the animals so that we could have a relationship with You. Amen.

Ref: Bower, Bruce. 1992. "Erectus unhinged." *Science News*, Vol. 141, June 20, p. 408.

Dusty Stars

Genesis 1:16
"And God made two great lights; the greater light to rule the day, and the lesser light to rule the night: he made the stars also."

The news has again been filled with stories about the search for planets orbiting nearby stars. All these stories assume that our Earth was not formed by God. They perpetuate the idea that the Earth and other planets in the solar system formed when dust around the young sun began to collect into lumps.

The search for planets orbiting other stars concentrates on stars that have dust around them. Scientists who believe in evolution think that if they find evidence of planets forming around those stars, evolution will have a stronger case against the Bible.

Research papers delivered at the American Astronomical Society discussed the dust around the star Beta Pictoris. They say that this star, visible from Earth's southern hemisphere, has chemical elements in its surrounding dust that are similar to the chemical elements they think were in the original dust cloud around our sun. Another paper said that the gaps in the dust surrounding eight other stars are due to the formation of planets around those stars. The distance of all these stars is too great to allow actual sightings of any orbiting planets. As David Black, director of the Lunar and Planetary Institute in Houston, correctly pointed out, "The landscape of previous efforts to detect planets is littered with the corpses of claims that haven't been substantiated."

Are there planets around other stars? We don't know. But we do know that the Bible's account of how the Earth was created by God is perfectly accurate.

Prayer: Dear Father in heaven, I give You thanks for the wonderful Earth You have created for us. Help those of us who know You to make good use of the knowledge we have about other stars and planets in our witness to others about You. In Jesus' Name. Amen.

Ref: Siegel, Lee. 1992. "Search for distant planets suggests there may be some." *The Beacon Journal* (Akron, OH), June 12, p. A6.

Giant Bacteria

Genesis 1:31
"And God saw every thing that he had made, and, behold, it was very good. And the evening and the morning were the sixth day."

Microbes. The very word means tiny. Bacteria are so tiny that we never knew about them until the microscope was invented.

Until recently, bacteria were thought to be just simple bags of living stuff. Anything that small had to be simple, right? With the more powerful microscopes of the twentieth century, scientists began to discover that there are many structures and organs inside a bacterium. However, it was still considered all right to talk about bacteria as "relatively simple."

Now scientists have discovered a giant bacterium that is changing the way they think about life. The newly discovered giant bacterium is a million times larger than the average bacteria. These creatures are so large that they can be seen without a microscope, although they appear only as little specks. The bacterium was originally discovered in the intestines of a surgeonfish. Since then, the creatures have also been found in mice and guinea pigs. Researchers say that the bacterium seems to be harmless, but scientists don't know what purpose it serves. Scientists also admit that if bacteria can become this large, the creatures are obviously more complex than science has suspected.

Those who believe in creation have been saying that there is no such thing as a simple form of life. Every living thing, no matter how tiny, no matter how large, has been specially designed by the Creator. And there is at least one purpose we know is served by this giant bacterium. It has taught unbelieving scientists that even bacteria are not as simple as they thought.

Prayer: I thank You, Lord, for the witnesses You have placed in our creation that point to You. Make me more effective in telling those whom You lead to ask about the forgiveness of sins that You have won for them. Amen.

Ref: "Chemical Forces Water to Run Uphill." *The Plain Dealer*, Sunday, June 21, 1992.

God's Safety Valve?

Psalm 71:6
"By thee have I been holden up from the womb: thou art he that took me out of my mother's bowels: my praise shall be continually of thee."

More than 75 percent of all women in their first trimester of pregnancy suffer some form of morning sickness. More than half become physically sick. While it may not help a woman with morning sickness feel better, it might help to know that morning sickness may serve a good purpose.

A University of California biologist has concluded that morning sickness may actually be the body's way of protecting the developing child. Margie Profet spent six years studying diets, birth defects, and the level of natural toxins in our food. Most food has a natural, low concentration of poisons. For example, plants make poisons to protect their leaves from marauding insects. Normally these poisons are so weak that they are completely harmless to us. However, the newly developing infant is extremely susceptible to these low levels of toxins until about the eighth week of development. Naturally occurring chemicals in our environment can cause birth defects and even death to the vulnerable infant. When the body detects a level of toxins dangerous to the developing child, it may use morning sickness to rid the body of them.

Profet mentions other studies that showed that women who get nauseous and vomit during early pregnancy have lower rates of miscarriage than women who don't get sick.

The final scientific opinion isn't in yet. Morning sickness may indeed by God's safety valve to protect the sensitive, developing child from harm.

Prayer: I thank You, dear Father, for the wonderful way in which You formed me in my mother's womb. I realize that as I was formed, You hand-made me no less than You originally formed the first human being, Adam. In Jesus' Name. Amen.

Ref: "Study: morning sickness may protect fetus from toxins." Minneapolis *Star Tribune*, June 13, p. 7A.

The Tuatara

Psalm 27:10
"When my father and my mother forsake me, then the LORD will take me up."

What's green, has three eyes and is one of the rarest animals on Earth?

The answer: the tuatara. This reptile is found today only on a few small islands off New Zealand. Its home on the islands in the Bay of Plenty and Cook Strait has no mammals at all. The tuatara was once also found on New Zealand. When settlers introduced mammals, the tuatara became extinct, probably because it could not compete with the mammals.

The tuatara is a reptile that is the only surviving species in its order. That means it's so different from other creatures that science knows of no similar creatures living today. It grows to a length of about two feet. The chunky reptile has a dark olive green body with light-colored spots. Its eyes are like those of the cat, except for one thing – it has three of them. Besides the eyes you would expect on each side of its head, it has a third eye on top of its head. Scientists don't believe that this third eye actually works. That's because its structure does not seem to be that of a complete, working eye. This third eye does have a retina and other eye structures. A nocturnal animal, the tuatara eats insects, worms and small animals. It buries its 12 to 14 eggs in a shallow hole in the ground, where they take up to a year to hatch.

The tuatara is a special example of God's creativity and His care for His creatures. It certainly isn't the fittest creature, yet God seems to have taken special care to preserve this unique animal.

Prayer: I thank You, dear Father, that Your love and care are handed out based on Your mercy rather than on human values. I thank You that for that reason You sent Your Son to pay the penalty for my sins so that You could forgive me. In Jesus' Name. Amen.

Lighter-Than-Air Food?

Isaiah 40:28
"Hast thou not known? hast thou not heard, that the everlasting God, the LORD, the Creator of the ends of the earth, fainteth not neither is weary? There is no searching of his understanding."

For years we have heard that seaweed would someday be the source of revolutionary products. Until recently, seaweed was known mainly for fouling boats and stocking health food stores with products that didn't catch on with the general public.

Now scientists at Lawrence Livermore National Laboratory have made a remarkable material from kelp. The solid material can be manufactured in a form that is so light that it weighs ten percent less than air. This form would float off in the same way that a helium-filled balloon does were it not for the fact that it has air bubbles trapped in it. Manufactured in a more dense form, it can support thousands of times its own weight. And you can eat it!

Called SEAgel, it's made from a commercially available substance called agarose that comes from kelp. Agarose is widely used as a thickener in foods. Scientists dissolve the agarose to make a gel. Then it's freeze dried. This simple and inexpensive process results in SEAgel. Scientists say SEAgel is an excellent insulator for aircraft, refrigerators, and oil tankers. It also makes a better packing material than foam chips. And it's biodegradable. It may also be used as a timed-release package for medicine and agriculture.

The discovery of how to make a lighter-than-air solid illustrates that modern science still has many things to learn about God's creation. Even modern science hasn't yet learned all the simple things. It has no business, then, challenging God's Word that He alone is the Creator.

Prayer: Dear Lord, help me to remember that we still have simple things to learn about Your creation when I hear the false pride of man claim that science knows so much that it can reject Your Word. Amen.

Ref: "From Seaweed, a Lighter-Than-Air Solid." *Science News*, July 4, 1992, p. 7.

Bird Barks and Lizard Growls

Numbers 22:30
"And the ass said unto Balaam, Am not I your ass, upon which you have ridden ever since I was thine unto this day? was I ever wont to do so unto thee? And he said, Nay."

Have you ever noticed that we can usually sense whether an animal is hostile or friendly, simply by its sound? Did you ever wonder about the universal features that allow for important basic information to be shared between animals and humans?

None of us ever had to be taught that a dog, growling deeply and showing its teeth, was trying to threaten us. We're all familiar with the high-pitched yips that same dog makes to welcome its owner home. Research shows that the motivations expressed by high-pitched and low-pitched sounds are universal among most creatures.

Researchers say that almost all animals bark (the high-pitched sounds) or growl (the low-pitched sounds). They were surprised to learn that animals don't simply make arbitrary noises. Sound has meaning. Even the higher-pitched voice of birds has a barking mode that can be seen on a graph and heard when a slowed-down recording is played. Animals make lower, harsher sounds when they're being aggressive. This is the growl. When friendly, an animal makes higher-pitched sounds. The bark seems to mean that an animal is neither hostile nor friendly, simply curious. Human speech follows the same general pattern.

These universal features of communication reflect the work of our Creator who intended for many different kinds of creatures to coexist. He gave us all a universal method for understanding important basic messages like fear, aggression and joy.

Prayer: I thank You, Lord, for the joy that animals add to our lives. I also thank You that You have made us able to share some basic but important features of communication. However, help me to value most highly Your communication to me in Your Holy Word. Amen.

Ref: Bennett, Dawn D. 1985. "Making sense of animal sounds." *Science News*, Vol. 127, May 18, p. 314.

Bee Cologne

Isaiah 53:5
"But he was wounded for our transgressions, he was bruised for our iniquities: the chastisement of our peace was upon him; and with his stripes we are healed."

Imagine being at a party where everyone's packed into one room and everyone is talking. Under these conditions it's often difficult to hear, or at least, understand what others around you are saying. Now imagine that all of these people look nearly identical. None has any noticeable distinguishing features that make him different from anyone else in the room. Now, to make matters worse, you have been informed that you are to find the person in the room who was not invited to the party.

That's the impossible situation that honeybees constantly face. But the noisy room I have just described is noisy in two ways. Bees not only communicate through sound. They also use scents to communicate. And the hive is full of scents. Yet honeybees are constantly alert to bees who are not members of the hive. Invaders are killed. How can bees so successfully protect the hive?

Entomologists have discovered that members of a hive all carry the same chemical password. The scent must be very distinctive because the hive is filled with smells. It also seems that all honeybees may use a specific combination and subtle variations of only two chemicals. Yet each hive can instantly tell whether another bee carries the specific chemical password.

The complexity of this avenue of beehive defense shows not only a wise and powerful Creator. It also shows us that He cares for what He has created. He cares for You, too. Learn more about the specific details of His love for You in the Bible.

Prayer: I thank You, Lord, that You have loved me so much that You carried the punishment for my sin in Your own body on the cross. Assure me that my sins are forgiven with Your peace, and grant me a renewed hunger to learn Your Word. Amen.

Ref: Pennisi. E. "Bees use Chemical Password to Show Kinship." *Science News*, July 4, 1992, p. 7.

Monkey Talk

1 Samuel 2:3
"Talk no more so exceeding proudly; let not arrogancy come out of your mouth: for the LORD is a God of knowledge, and by him actions are weighed."

Several years ago, Koko the gorilla amazed scientists and the public by apparently learning enough of the rudiments of English to communicate simple thoughts. Now studies on wild vervet monkeys suggest that Koko's achievement might be nothing more than a normal, God-given ability.

Anthropologists from UCLA report that wild vervet monkey communication shows much more language development than scientists ever expected. It took years of study in the wild before anthropologists began to hear subtle differences in the grunts and screeches offered by the monkeys. They confirmed these differences by recording the different sounds and the actions that went with the sounds. Then they analyzed the sounds electronically and found that there was indeed a pattern of differences in the sounds that corresponds to specific actions.

Anthropologists then set up their sound equipment in the monkey's home territory to test their interpretation of the sounds. When playing the grunt that researchers thought meant "open plain", monkey listeners looked out toward the open plain. Likewise, when the "other group" grunt was played, the monkeys searched the open plain even more carefully. The scientists had discovered that they had successfully understood two other monkey phrases as well.

These vervet monkeys teach us that language is not the evolutionary development of early humans, but a gift from God, the Author of language.

Prayer: Heavenly Father, I thank You for the gift of language. Help me to make better use of this gift in praising You and telling others of Your love for us through the forgiveness of sins through Jesus Christ. Amen.

Ref: Greenberg, John. 1985. "The sophisticated sounds of Simians." *Science News*, Vol. 127, June 15, p. 356.

Hard-Working Virus

Philippians 1:22-24
"But if I live in the flesh, this is the fruit of my labor: yet what I shall choose I wot not. For I am in a strait betwixt two, having a desire to depart, and to be with Christ; which is far better: Nevertheless to abide in the flesh is more needful for you."

When it was introduced, some people thought that it was against God's will. Since the human body was made in God's image, they thought it was wrong to put something into the human body that was made from a cow. An artist of the day even drew a cartoon showing people receiving the vaccine that was made from a cow. It showed cows emerging from the leg of someone who had been vaccinated. The vaccine was smallpox vaccine, and the year was 1796.

The first smallpox vaccine worked, and still works, very nicely. It has eradicated smallpox. The weakened virus used in the vaccine is called vaccina, and it is still hard at work today.

Viruses do their damage by fooling the cell into letting it latch on. It then takes over, using its genetic information and the cell's machinery to turn out its own proteins to make more viruses. The effect on us is the disease characteristics of the infecting virus. Vaccina, however, is weakened, so the body's immune system can kill it without allowing the disease to develop. The body remembers that immune response and fights the disease more successfully the next time it shows up. Modern genetic engineers are adding the disease-causing genes of many viruses to the vaccina virus. In animal tests, the new vaccina virus has prevented malaria, hepatitis B. herpes simplex, rabies, and other diseases.

Of course, the image of God does not refer to the physical form of the human body. Vaccines have been a great blessing, saving the lives of many so that they, too, might serve God.

Prayer: Dear Father, I thank You for the blessings that we enjoy because of vaccines. At the same time, do not let us forget that our greatest need for rescue is not from diseases of the body, but from sin – a rescue You have provided through the forgiveness of sins, won for us by Christ. In His Name. Amen.

Ref: Miller, Julie Ann. 1985. "A vaccine for all seasons." *Science News*, Vol. 127, June 15, p. 379.

Smarter Than They Thought

Ephesians 6:4
"And, ye fathers, provoke not your children to wrath: but bring them up in the nurture and admonition of the Lord."

Traditional approaches to early childhood development have relied heavily on the claims of those who don't believe in the Creator. As a result, these approaches see human beings as simply another animal. Along with this, they usually think that humans are born into the world as a blank slate with no knowledge. As a result, the traditional view has assumed that everything an infant does is based on instinct.

Few have ever tested these beliefs scientifically. Since the viewpoint agrees with evolution, theorists figured they must be correct. At the same time, many Christians simply accepted the claims because they thought they had been tested and proved scientifically.

Evolutionary thinking led Swiss psychologist Jean Piaget to say that infants are not able to imitate facial gestures until they are about a year old. He further said that infants are not able to imitate actions they have seen before until they are 18 months to two years old. It wasn't until the last few years that these claims were tested. Researchers found that infants were much smarter than they ever imagined. Infants only 72 hours old – in one case, only 42 minutes old – will imitate someone sticking out his or her tongue. Infants as young as 14 months will imitate actions they saw the day before.

Researchers concluded that infants are born with eagerness and ability to begin to learn from adults. This simple truth makes our example as Christians important to even the youngest child.

Prayer: Help me, Lord, to be a good example to everyone around me, especially the youngest children. Help me to identify and rid myself of ideas about very young children that may have been influenced by the unbelieving world, and to think of children more in the way that You do. Amen.

Ref: Miller, Julie Ann. 1985. "Born smart: imitation of life." *Science News*, Vol. 127, June 15, p. 376.

The Spiny Postage Stamp

Genesis 1:21
"And God created great whales, and every living creature that moveth, which the waters brought forth abundantly, after their kind, and every winged fowl, after his kind: and God saw that it was good."

After more than a century of digging Earth's hills, valleys and plains, paleontologists have uncovered and catalogued more than a billion fossils. Evidence has been found of many plants and animals that are extinct today. No evidence has ever been found showing new plants or animals developing, however.

New creatures discovered for the first time in the fossil record normally fit into a known phylum. Recently, paleontologists working in northern Greenland announced their discovery of a strange, extinct creature. The five-and-a-half-inch-long creature is so strange that it does not seem to fit into any known phylum. Paleontologists have given the fellow the name "Grasper" because it has a pair of appendages sticking out the front that may have been used to grasp things. The body has been described as being shaped something like a spiny postage stamp. It had gill-like flaps along each side and bumps along the back. Grasper also had a set of spines sticking out from its back. Only one complete skeleton of the creature has been discovered, but scientists have about 50 partial skeletons.

Paleontologists also report a number of other strange animals at the site that do not fit into any phylum. One looks like a swimming vacuum cleaner.

The fossil record offers us some surprising examples of God's creativity. The fossil record doesn't offer any evidence that the creatures alive today developed from other creatures.

Prayer: Father in heaven, give Your people understanding. Help us not to fear what we might learn from new knowledge, but help us learn how to sort true knowledge from interpretations offered by those who reject Your truth. In Jesus' Name. Amen.

Ref: Monastersky, Ronald. 1992. "Evolutionary oddball surfaces in Greenland." *Science News*, Vol. 142, July 11, p. 22.

The Wandering Compass

Job 26:7
"He stretcheth out the north over the empty place, and hangeth the earth upon nothing."

The Earth's magnetic field makes compasses work so that Boy Scouts don't get lost in the woods and ships arrive at the correct port. It actually moves over time. This means that navigators periodically need updated information on the position of the pole for their charts.

Many people have the idea that the Earth's magnetic field moves very slowly. The Earth's magnetic field actually offers some of the clearest evidences that the Earth is young. Geophysicists recently reported that the north magnetic pole has moved 800 kilometers northwest of where it was in 1904. That's a much faster rate of movement than many people imagined. Its speed and direction of movement are about what was expected by geophysicists, however. They also report that the magnetic pole has moved as much as 80 kilometers in one day.

Measurements taken over the last 150 years show that the Earth's magnetic field is decaying. When Christ walked the hills of Palestine, the magnetic field was twice as powerful as it is today.

The Earth is much more dynamic and changeable than most people, including many scientists, think. These rapid changes in the position of the magnetic pole and the rapid decay of the magnetic field offer evidence that the Earth is much younger than many people think. One of the world's best-known experts has shown that the decaying magnetic field means that the Earth could not be much older than the Bible's genealogies indicate.

Prayer: Lord, the blessings of modern travel rely on Your gift to us of the Earth's magnetic field. I also thank You for the beauty of the northern and southern lights that is created by the magnetic field. Amen.

Ref: "News of Magnetic Poles." *Science News*, Vol. 128, p. 8.

Physician Wasp?

Job 36:5
"Behold, God is mighty, and despiseth not any: he is mighty in strength and wisdom."

In previous "Creation Moments" features we've talked about wasps that control crop pests. Natural controls – such as these wasps – show the Creator's hand in designing the creation. New facts about how wasps control pests make our Creator's hand even more evident.

There are several thousand species of wasps that lay their eggs inside crop pests or use the pests for food. These wasps have become a popular form of natural pest control among farmers and gardeners.

Researchers have now discovered that the wasps do more than simply feed on the pests or use them as food for their young. Typically, the wasps will inject their eggs into worm-like pest larvae. It has been discovered that the wasp eggs are coated with a virus that holds the pest larvae in the immature stage until the developing young wasps have no more need of the egg. The virus appears to move to an organ inside the pest larva and affects the insect's immune system. It also acts on the larva's endocrine system to take away the larva's appetite. So the pest larva destroys less crop and, being starved for food, fails to develop into an adult. This gives the wasp eggs plenty of time to develop into young wasps who will finish off the pest.

Researchers are still trying to discover exactly how the virus's carefully orchestrated attack strategy works. Clearly, no creature, nor the creation itself, invented this strategy. It could only have been conceived and built by a Master Biochemist and Physician!

> ***Prayer：I thank You, Lord, that You have built into Your creation natural ways to control creatures that can become unwanted pests. Help us to learn more about the natural controls You have made so that we will be better able to feed a growing population. And with that food, let Your people bring others Your saving Word. Amen.***

Ref: Raloff, Janet. 1985. "Virus allows wasps to kill crop pests." *Science News*, Vol. 128, p. 22.

Bacteria That Dissolve Steel

Matthew 6:19
"Lay not up for yourselves treasures upon earth, where moth and rust doth corrupt, and where thieves break through and steal:"

They work silently and in the dark. While most of Earth's inhabitants need oxygen for life, they merely tolerate it. They prefer to build their own environments where there's no oxygen. Then they go to work. And they love metal. Using complex chemistry, they begin dissolving metal. The can make a sixteenth-of-an-inch hole through an inch-thick pipe in six months. Stainless steel isn't so tough – it doesn't slow them down a bit. Even modern space-age metals like titanium can't stand up to them.

No, we're not talking about some hideous creatures from outer space. These strange-sounding creatures are called sulfate-reducing bacteria. Each year, metal corrosion causes about $167 billion in damage. And a large part of that damage is caused by bacteria destroying metal pipes.

Sulfate-reducing bacteria begin by sealing off their colony from liquid in a pipe or tank. Once sealed off, bacteria begin forming hydrogen gas. Sealed under the biosphere, the hydrogen accumulates and is absorbed by the metal. The absorbed hydrogen begins to corrode and make the metal brittle. Researchers using pipes coated with epoxy couldn't stop the bugs. It appears that the bugs thought the pigment in epoxy made a great change of diet.

Despite our modern scientific sophistication, moth and rust continue to afflict our efforts. This is God's way of reminding us that we, along with the creation, are afflicted with sin and in need of the forgiveness of sins that is ours only through Jesus Christ.

Prayer: I pray, Lord, that You would keep me mindful of the vanity of placing my trust in my own or others' efforts. I know that ultimately You are my only hope for meaning here and in eternity. Forgive me for those times I have trusted in my own wisdom and efforts, and grant me Your peace. Amen.

Ref: Raloff, Janet. 1985. "The bugs of rust." *Science News*, Vol. 128, July 20, p. 42.

The Young Lizard's Canteen

Isaiah 43:20
"The beast of the field shall honour me, the dragons and the owls: because I give waters in the wilderness, and rivers in the desert, to give drink to my people, my chosen."

When you go into the desert, it's always a good idea to take a canteen of water with you. Our Creator applied that same simple wisdom when he designed the Yarrow's spiny lizard.

Like many desert dwellers, Yarrow's spiny lizards are well designed to conserve water in the dry desert climate. For example, they don't even have a urinary bladder. Their bodies are designed to remove water from their waste before it is released. The water is recycled within the body, since water is a precious resource in the desert.

Unlike many reptiles, Yarrow's spiny lizard bears live young. The two-inch-long newborn lizards can dry out quickly in the desert. To increase their ability to survive, the Creator gave the newborns a built-in canteen of water. At birth, a full canteen makes up 10 percent of the lizard's weight. After the first month of life, the canteen shrivels and is no longer used. This, too, is a wise design, since the growing lizard has learned the ways of the desert and now has a greater need for speed than it has for a bulky water reservoir.

I don't believe that any of us can fully appreciate God's limitless imagination and creativity that we see in the living world around us. We can appreciate the fact that He is willing and able to provide practical and workable solutions to every problem that we face. His solutions, beginning with our greatest need – the forgiveness of sins – are found in the Bible. The trust to apply His solutions to our lives is found in faith, worked by His Holy Spirit.

Prayer: I thank You, Lord, for Your limitless love and patience with me. Forgive me for those times when I have sought my own solutions rather than Yours. Grant me Your Holy Spirit so that I may make better use of Your Word in my life. Amen.

Ref: "Canteen for a young lizard." *Science News*, Vol. 128, p. 124.

Alaskan Dinosaurs

Genesis 1:7
"And God made the firmament, and divided the waters which were under the firmament from the waters which were above the firmament: and it was so."

In Genesis 1:7, Scripture describes God's division of the waters. The creation account tells us that God used a "firmament" to divide the waters into those that were above the firmament and those that were below the firmament. Careful study of the original Hebrew indicates that the firmament was a relatively thin layer of something between the waters of the seas and waters above. When pictures of our Earth were first taken from space, they showed that the atmosphere is indeed a thin layer, hugging the surface of our planet.

Meteorologists who believe in creation have shown that a fine canopy of water vapor several miles above the Earth would have allowed the sun and stars to shine through. Yet, it would have created a greenhouse effect that kept the entire Earth's climate relatively even and mild. That fits with the geological evidence that the Earth once had a warmer and more even climate.

Dramatic scientific verification of this explanation came in 1961 when the bones of three dinosaur species and two tropical reptiles were discovered on the Alaskan tundra near Prudhoe Bay. Today, that area is far too cold to support such life. This dramatic find was not even revealed until 1984. Since then, other tropical remains, including more dinosaurs, have been discovered in the far north.

While we don't look to science to prove the Bible, we know that accurate scientific findings will never disagree with what the Bible tells us about Earth history.

Prayer: I thank You, dear Father, for giving us Your Word. While the main purpose of Your Word is to reveal salvation to us through Your Son, Jesus Christ, I thank You that Your Word is trustworthy in all that it says. In Jesus' Name. Amen.

Ref: Dusheck. J. "Arctic Dinosaurs Raise Questions." *Science News*, August, 1985, p. 135.

Talking About the Takin

Proverbs 3:5-6
"Trust in the LORD with all thine heart; and lean not unto thine own understanding. In all thy ways acknowledge him, and he shall direct thy paths."

In has a bulky, humped body like a bear. Its hind quarters slope downward like a hyena. It has legs like a cow. Its tail looks as if it were borrowed from a goat. It has the horns of a wildebeest and a face like a moose with mumps. Some scientists think the golden fleece Jason brought to Greece in classical times was from a takin. One scientist commented that if the camel was designed by committee, the same committee designed the takin from the parts that were left over.

The takin is a large hoofed animal that can weigh up to 650 pounds and stand five feet high at the shoulders. It lives in China's remote, rugged mountain forests, where it is protected. Takins are herding animals that eat plants. Scientists have noted over 100 species of plant that the takin eats. One scientist suggested that it might be easier to put together a list of the plants takins don't eat.

Standing on their hind legs, takins can reach branches eight feet high. They will push over thin trees to get at the leaves. And they're large enough to straddle a five-inch tree and bend it down until they can get at the leaves. Scientists were puzzled because for some time they never saw any young takins in the grazing herds. Then they discovered that the young are often tended in the takins' version of a day care center, while the rest of the herd goes off to socialize and feed.

Our Creator's imagination knows no limits. In its uniqueness, the takin glorifies our Creator God.

Prayer: Father, Your unlimited imagination has done things that we cannot imagine or understand. Help me to remember this fact when I am tempted to trust in those things that my limited imagination can understand in place of Your clear Word of truth. In Jesus' Name. Amen.

Ref: Miller, Julie Ann. 1985. "Observations on a legend: takin in the wild." *Science News*, v. 128, Aug. 31, p. 148.

Brain Disappoints Evolutionary Thoughts

Psalm 119:130
"The entrance of thy words giveth light; it giveth understanding unto the simple."

According to evolutionary thought, humans came from ape-like creatures. These ape-like creatures came from reptiles through many steps. Likewise, reptiles ultimately came from fish through many steps. Evolutionists say that with each major stage, more parts were added to the brain. The final development, and the greatest of all, is the cerebral cortex.

According to this theory, one of the oldest parts of the brain is the part deep inside. Evolutionists claim that this part of the brain, called the basal ganglia, comes from our reptilian past. They have described this part of the brain as "primitive." It controls such simple things as movement and spatial memory. These are basic functions needed by reptiles. Of course, the evolutionary view that the basal ganglia is primitive was never scientifically researched.

Then researchers started to examine the workings of the basal ganglia. They have concluded that the deep structures of the brain that are supposed to be primitive are actually quite sophisticated. In fact, these so-called reptilian structures rival the cerebral cortex in sophistication. These structures receive input from all parts of the cortex. And all the neurochemicals found anywhere else in the body are also found in these structures. In other words, there is absolutely no evidence of their being primitive.

Once again scientific research has not found what evolutionists expected. The human brain has not developed by adding parts to reptile brains. It was specially created just for human beings.

Prayer: *I thank You, dear Lord, for the wonderful abilities of my brain. Forgive me for those times when I have failed to use the abilities You have given me – especially when I have failed to use them in Your service. Amen.*

Ref: Miller, Julie Ann. 1985. "Deep core of brain gains respect." *Science News*, Vol. 128, Nov. 9, p. 297.

Touching Love

2 Corinthians 13:11
"Finally, brethren, farewell. Become perfect, be of good comfort, be of one mind, live in peace; and the God of love and peace shall be with you."

Love is one of the necessities of life that is both physical and spiritual. While it doesn't fit with a strictly materialistic view of life, love and touch, or lack of love and touch, can determine our health and sometimes even life and death.

The claim that love and touch are necessary to life and health was considered so outrageous that it wasn't until relatively recently that scientists began to test such claims. Researchers selected 40 hospitalized infants. These infants were not in intensive care, on oxygen, or intravenous feedings, but they were too tiny to leave the hospital. Researchers randomly selected half the group for attention. These infants were touched for three 15-minute periods a day. The first five minutes were spent stroking the infants' bodies. The next five minutes were spent lightly moving the infants' arms and legs. This was followed by five more minutes of stroking.

Even though the infants often didn't seem to notice the attention, it made a difference. After only 10 weekdays of this attention, the stroked infants had gained 47 percent more weight than the infants that were ignored. This is especially amazing considering the infants that gained less weight had been fed the same number of calories. The stroked infants were also more active, so they should have actually metabolized more of the calories and lost weight rather than increasing in weight.

Can we literally live on love? Well, we do need food and other things, but we also need love – especially from our Creator.

Prayer: Father in heaven, I thank You that You have loved me and sent Your Son, Jesus Christ, to carry my burden of sin and guilt on the cross. Never let me doubt Your forgiveness, and continually assure me of Your forgiving love – for I know that this is where peace is found. In Jesus' Name. Amen.

Ref: Bower, Bruce. 1985. "Different strokes." *Science News*, Vol. 128, Nov. 9, p. 301.

Science Confronts Meditation Claims

Psalm 119:148
"Mine eyes prevent the night watches, that I might meditate in thy word."

Many people promote meditation as harmless and healthful. Of course, growing interest in meditation cannot be separated from the increased popularity of Eastern and New Age religion.

Many don't place the claims that Eastern meditation reduces stress in the same category as the magical New Age claims made for crystals. Unfortunately, even children in public schools go through Eastern meditation exercises.

Eastern meditation has often been recommended as one treatment for hypertension. However, no one had ever studied meditation's medical claims to see if they were true. Then the results of such a study were published in the *Journal of the American Medical Association*. Government researchers had studied effects of Eastern meditation on more than 2,100 men and women with high blood pressure. They concluded that it is ineffective in lowering blood pressure. Yes, it is true that there are changes in the brain waves of someone who is meditating. Brain waves normally change when one type of mental activity replaces another. Nor is Eastern meditation a particularly good choice to produce brain waves associated with mental peace.

Eastern meditation is not harmless. It can open the mind to negative influences. And claims that it can reduce the effects of stress have been scientifically shown to be untrue. Let's return to our Bible and say with the Psalmist, "My eyes are awake through the night watches, that I may meditate on Your word."

Prayer: I thank You, dear Father, that You have given us Your sure Word in Scripture and the Word made flesh for our salvation. Increase my eagerness to meditate on Your Word. In Jesus' Name. Amen.

Ref: "20 medical stories you may have missed." *U.S. News & World Report*, Aug. 3, 1992, p. 58.

Frog Pharmacists

Matthew 13:22
"He also that received seed among the thorns is he that heareth the word; and the care of this world, and the deceitfulness of riches, choke the word, and he becometh unfruitful."

Some frogs are regular little drug factories. Their skins produce a wide range of powerful alkaloids, a large family of chemicals usually produced by plants. Familiar alkaloids include quinine, caffeine and morphine. Different alkaloids have different effects, usually on the nervous system.

The poison-arrow frog may be the most famous alkaloid-producing frog. South American natives simply rub their arrows or darts on the frog's back. One species of this frog contains enough poison to kill about 100 people! Another frog, native to Ecuador, produces an anesthetic more powerful than morphine. South American natives rub another frog on their wounds. Researchers discovered that the frog's skin makes a powerful antiseptic. Yet another frog alkaloid will change the colors on parrot feathers. Some of the alkaloids produced by frogs are so complex that it takes years of study, using modern science's most sophisticated equipment, to discover the active chemical.

Researchers in this field are only beginning to learn about the alkaloids produced by frogs. They have yet to research a frog that natives claim to use to heighten their senses for the hunt. What puzzles scientists who believe in evolution is that the frogs' exotic skin chemistries don't follow the patterns they anticipated based upon expected evolutionary relationships.

This means that these frogs have one more exciting purpose. They provide evidence against evolution and for the Creator.

Prayer: I thank You, Lord, that You have included chemicals in the creation that can help us treat our medical problems. I ask that I would not become so concerned with earthly cares and worries that I forget about my spiritual needs – which are fully supplies by You. Amen.

Ref: Pennisi, Elizabeth. 1992. "Pharming frogs." *Science News*, Vol. 142, July 18, p. 40.

Raccoon Bear?

Romans 1:24-25
"Wherefore God also gave them up to uncleanness through the lusts of their own hearts, to dishonor their own bodies between themselves: Who changed the truth of God into a lie, and worshipped and served the creature more than the Creator, who is blessed for ever. Amen."

In creation, things aren't always as simple as they seem. Even before Charles Darwin, scientists who believed in evolution classified plants and animals by appearance. A story about the creature's evolutionary history would be invented, based on its appearance and, sometimes, its habits. God is too creative, however, for this kind of invention to write Him out of history.

This background information helps us understand the problem of the giant and lesser pandas. Are they related to bears? Are they related to raccoons? Or neither? Evolutionary scientists have been debating this for more than a century. They were sure the giant and lesser pandas should be grouped together because of similarities in their teeth. And even though the giant panda looks like a bear, it doesn't hibernate. Nor does it roar – it has a very unbear-like bleat, like a lamb.

Modern molecular biology has now investigated the panda question. DNA and proteins from greater and lesser pandas were examined and compared. Researchers arrived at a startling conclusion, based on what they knew about raccoon and bear DNA and proteins. The greater panda is a bear. The lesser panda is not a bear, but is related to the raccoon.

In their rush to deny the Creator, humans build huge claims out of tiny amounts of information. It is, in fact, much easier to see the greater and lesser pandas as unique creatures that have come from the hand of our imaginative Creator.

Prayer：I thank You, Lord, for the wonderful creatures we call pandas. In addition, I thank You that they witness to Your creativity and confound those who wish to deny You. Protect these creatures so that we may always enjoy them. Amen.

Ref: "Panda pedigree: giant and lesser." *Science News*, Vol. 128, p. 216.

God's Mind, Human Mind

Isaiah 55:8
"For my thoughts are not your thoughts, neither are your ways my ways, saith the LORD."

It was only a few years ago that excited supporters of the big bang theory for the origin of the universe announced that their theory had apparently been proven. Now their unverified conclusions are being seriously questioned on several fronts.

Even as it was being announced that the big bang theory was proven, verification was being made on other results showing that the big bang theory is probably wrong. William G. Tifft of the University of Arizona had been looking at lists of red shifts from distant galaxies. On a graph, he plotted the red shifts corrected for the effects of the Earth's motion. If the red shifts from these objects are the result of their speeding away from Earth, their red shifts should be evenly spread over the known range of shifts. Tifft found that they're not. The red shifts are clumped into patterns. The intervals are about 72 kilometers per second, with some half that value and some one-third of that value. Astronomers from the Royal Observatory in Edinburgh verified his results.

A second study by Tifft has shown that galaxy red shifts measured from Earth have changed in just a few years! This change is completely unexpected by proponents of the big bang theory. The change is large enough that Tifft expects to be able to test the rate at which these changes take place within only a few years! One suggestion is that red shifts may be some sort of decay phenomenon rather than expansion of the universe.

How could human beings expect to understand a universe created by the Author of Scripture when they reject Scripture itself?

Prayer: I thank You, dear Lord, for the revelation of Your Word of salvation in Scripture. I pray that I may never presume to understand something before I know and believe what You have said about it in Your inerrant Word. Amen.

The Not-So-Gentle Whale

Genesis 1:26
"And God said, Let us make man in our image, after our likeness: and let them have dominion over the fish of the sea, and over the fowl of the air, and over the cattle, and over all the earth, and over every creeping thing that creepeth upon the earth."

Like dinosaurs, whales are huge and wondrous creatures whose sheer size and majesty glorify the Creator. Some whales are even larger than were any of the dinosaurs.

Whales were hunted for generations for processing into a variety of products. It was not until they were hunted to near extinction that governments began to work to prevent this from happening. Much of the hunting was justified on the grounds that whales were ruthless killers. After that perception of whales was proven wrong, the pendulum swung the other way. As a result, the public, and even many scientists, began to romanticize whales. They were presented to the public as gentle, benevolent giants. It became stylish to talk about the whale as an intelligent creature, rivaling and perhaps surpassing humans. Now more objective research into whale behavior shows that whales, while magnificent creatures, are nothing more than wild animals.

Researchers have learned that while whales can be gentle, they can also be terribly violent. They have documented males in vicious battles for the attentions of females. Sometimes as many as ten males will be involved in a free-for-all. Their fighting tactics include ramming each other at full force with their 40-ton bodies. Whales are not very loyal either. Males will mate with several females. Worse, they will attack females to force them to mate.

Unlike humans, whales are animals, and only humans were created to have an intimate relationship with their Creator.

Prayer: *I thank You, dear heavenly Father, for making those wondrous creatures, the whales. Help my life to show that I was truly made for an intimate relationship with You through the forgiveness of sins that has been earned for me through the innocent suffering and death of Your Son, Jesus Christ. Amen.*

Ref: Carpenter, Betsy, with Karen F. Schmidt. 1992. "Whales." *U.S. News & World Report,* July 13, p. 58.

Scientist Sea Lions

Job 12:9-10
"Who knoweth not in all these that the hand of the LORD hath wrought this? In whose hand is the soul of every living thing, and the breath of all mankind."

Does a sea lion make a good scientist? Human scientists say that sea lions show a good understanding of complex relationships between variables.

At the Long Marine Lab at the University of California at Santa Cruz, sea lions are indeed in training for scientific careers. The sea lions will not only be doing research on themselves, but also on other sea creatures. One of their most interesting tasks is videotaping gray whales in the ocean depths. This past summer, they were training in tanks to learn how to take good videos. Scientists feel that since the whales are used to seeing sea lions, they will behave more naturally than they do when human beings are taping them. In addition, sea lions can dive to depths that would be difficult for human divers to reach. The sea lions will also wear monitors that record their own body functions.

Trainers describe sea lions as more intelligent than dogs. Sea lions prefer interaction with humans, especially games, to food as a reward. One researcher noted that the sea lions now undergoing training will select a Wiffle ball rather than fish, when given a choice. Scientists say that the sea lions in training will follow a broad range of instructions if they are kept entertained. They will dive to specific depths and remain there until instructed to surface.

God has given His creatures intelligence based on their needs, not on their resemblance to man. This simple principle explains why the claims that man arose from the animal world are wrong.

Prayer: I thank You, dear Lord, for the wonderful creatures You have made, especially those that interact with us. Help those who work with the creatures You have made and clearly see how these creatures glorify You and proclaim Your work of creation. Amen.

Ref: Angier, Natalie. "Getting a Handle on the Sea Lion's Share of the Work." *Star Tribune*, Wed. July 15, 1992.

Secret Intelligence from the Deity

Genesis 1:21
"And God created great whales, and every living creaure that moveth, which the waters brought forth abundantly, after their kind, and every winged fowl, after his kind: and God saw that it was good."

Whales can maneuver precisely through the world's oceans and ice floes and among each other. Herman Melville noted this when he wrote in Moby Dick that whales seem to be guided by a "secret intelligence from the Deity."

Much about that "secret intelligence" remains a secret to this day. It is known that sperm whales can echolocate in the same way that bats do. They send out a series of clicks from a special organ in their head. Other species of whales, however, lack that organ. Scientists think that other whales may use a less precise method of echolocation based on their low-pitched calls.

Whales also use sounds to communicate with each other. Scientists have catalogued at least 23 different sperm whale click arrangements. Some of the patterns always follow other patterns, indicating that the clicks are a form of language. Some whales are believed to carry out conversations over thousands of miles. Scientists believe that some of the frequencies used by the whales allow them to communicate with other whales halfway around the world. The haunting melodies of the humpback whale are mating songs. The songs can last as long as 30 minutes. A humpback song will be the same within a single population of whales and sometimes within an entire ocean basin!

Whales are beautiful creatures. While they are intelligent, they are not as intelligent as once thought. However, they still glorify their Creator.

Prayer: Dear heavenly Father, I thank You for making whales and showing forth Your glory and Godhead by giving them such wonderful abilities. Help me to use the abilities You have given me to glorify You, too. In Jesus' Name. Amen.

Ref: Carpenter, Betsy, with Karen F. Schmidt. 1992. "Whales." *U.S. News & World Report*, July 13, p. 58.

Damascus Steel

2 Samuel 22:28
"And the afflicted people thou wilt save: but thine eyes are upon the haughty, that thou mayest bring them down."

A new superplastic steel might be the rediscovery of a 2,000-year-old steel-making process.

Normally, steel that is easy to pour into a mold is also softer. Parts made from it wear more easily. Tougher steels are difficult to mold and need a great deal of machining after being molded. This additional step accounts for 30 percent of the price of a steel object. Now a new mixture of steel has been developed that molds more easily than other steels, and yet is more resistant to wear. The new steel can flow like taffy into a perfect mirror image of the mold. No additional machining is necessary after molding. When subjected to the right combination of heating and working, conventional steels can only be stretched eight percent. The new steel can be stretched 1,100 percent!

Some scientists think that the new steel is the famed Damascus steel that was used to make swords and knives almost 2,000 years ago. The formula for making Damascus steel has been lost. Existing records of how the steel was made are not very helpful. They advise steel makers to add milkweed to the steel mixture. Others dispute the claim that the new steel is Damascus steel because the new steel lacks the intricate pattern that appears on Damascus steel surfaces.

While our twentieth-century knowledge is impressive, we do well to keep it in the right perspective. Rather than seeing man on an upward climb, we should remember that it is entirely possible that throughout our history we have forgotten more than we know today.

Prayer: *You, dear heavenly Father, are the source of all true knowledge and wisdom. Help me to remember this simple truth in a world that seems to have lost this basic conviction. In Jesus' Name. Amen.*

Ref: Donald Woutat. "New Steel may be Stuff of Legend." *Star Tribune*, Tuesday, June 23, 1992.

The Talking Ear

James 1:22
"But be ye doers of the word, and not hearers only, deceiving your own selves."

Can your ear talk to you? New research seems to show that your ear actually generates sound that echoes whatever you hear.

Startled scientists are making new discoveries that appear to show that we have hearing aids built into our ears. Yes, our ears do produce sounds. When sound hits the eardrums, the vibrations move bones within the ear, causing a bone called the stapes to vibrate. These vibrations are translated to pneumatic pressure within the cochlea. The oscillating pressure is picked up by tiny inner hairs in the cochlea that vibrate with the sound. These vibrations generate an electrical signal that is sent to the brain.

Researchers have learned that outer hair cells within the cochlea respond to the incoming signals by generating audible sounds that can be picked up by tiny microphones. The sounds are generated as the hairs dance up and down in time with the incoming sound, just like the cone of a loudspeaker. The effect is that the ear echoes the incoming sound a few thousandths of a second after it enters the ear. The echo generated within the cochlea is not necessary for hearing. However, scientists suspect that this feedback system helps people smoothly hear sounds that range from soft to loud. Some scientists have voiced their skepticism about these astonishing findings.

Our ability to hear sound is much more elegant in design than scientists ever expected. The technically precise details of the ear's design discredit all claims that the ear could have evolved.

> **Prayer: Lord, let my voice praise You for all the great things You have done. Let me always be glad to hear Your Word and, assured of Your forgiving grace, put it into practice in my life. Amen.**

Ref: Malcolm M. Browne. "Let's Hear it from the Ears." *The Plain Dealer*, Sunday, June 21, 1992.

The Messiah of the Dead Sea Scrolls

Acts 17:2-3
"And Paul, as his manner was, went in unto them, and three Sabbaths days reasoned with them out of the scriptures, Opening and alleging, that Christ must needs have suffered, and risen again from the dead; and that this Jesus whom I preach unto you, is Christ."

A debate has broken out over the image of the Messiah that's presented in one of the Dead Sea Scrolls. Two American scholars have claimed that one of the scrolls talks about the leader of the religious community being put to death.

Since the scrolls date from about 200 B.C. to 50 A.D., some scholars say that this means the idea of an executed leader is not unique to Christianity. Modern biblical critics view Christianity as a result of human social evolution rather than revelation. Therefore, they try to find the evolutionary steps in the development of Christianity. According to their translation, the scroll in question reads: "and they put to death the leader of the community, the Branch of David, with wounds (also "stripes" or "piercings").

However, Hebrew scholars convened a special seminar that brought together 20 scholars from around the world to study the question. They unanimously concluded that the original translation, offered by an Oxford scholar, is accurate. The correct translation says that the Branch of David will kill Israel's enemies, not be killed by them. They base this conclusion not only on the Hebrew, but on several other texts that speak of the Messiah as a leader who shall free Israel from her political oppressors. This was, in fact, the common expectation at the time of Christ.

It is impossible to understand Christianity when one rejects divine revelation. Christianity is not the product of man's mind, but of God's love.

Prayer: Lord, I pray that You would make Your people wise and more knowledgeable so that they would not be easily fooled by those who do not accept Your Word. Amen.

Ref: "Scrolls' Messiah Reinterpreted." *Star Tribune*, Friday, July 10, 1992. "The 'Pierced Messiah' Text-An Interpretation Evaporates." *Biblical Archaeology Review*, July/Aug. 1992, pp. 80-82.

Misidentification

Galatians 1:8
"But though we, or an angel from heaven, preach any other gospel unto you than that which we have preached unto you, let him be accursed."

One of the best strategies for overcoming an opponent is to fool the opponent into thinking you do not exist. Then you infiltrate your opponent and make him think that you are on his side.

This is the tactic used by the microorganism responsible for the potentially fatal tropical disease, leishmaniasis. The disease is spread by a number of species of tropical blood-sucking sand flies. When they attack, they inject some of the microorganisms into the bloodstream. Infection can cause painful skin sores. If the protozoans migrate to internal organs, they can fatally injure the liver or the spleen. The result, then, is a fatal disease called the black sickness.

Normally, the organism that causes the disease would be consumed and destroyed by the white blood cells as part of the normal working of the immune system. However, the bugs actually live within the white blood cells that have consumed them and subvert the cells themselves. They take over a cells' chemical machinery and prompt it to make a growth factor called TGF-beta. This chemical neutralizes the chemical that white cells use to kill the microorganisms. It's normally made by the white cells only after an infection has been stopped. In other words, the proteins fool the immune system into giving the all-clear.

We are most spiritually vulnerable when we think that we have lived such good lives that we are in no need of having our sins forgiven. That lie subverts our dependence on our only Savior from sin, Jesus Christ, in Whom *only* we find forgiveness.

Prayer: Forgive me, Father in heaven, for those times when I have congratulated myself on my goodness, rather than relying on the only One who is perfect – my Lord and Savior, Jesus Christ. In His Name. Amen.

Ref: Ezzell, C. Disarming, "Combating a Tropical Parasite." *Science News*, July 1992, p. 53.

A Trail That Tells a Tale

Psalm 19:1
"The heavens declare the glory of God; and the firmament sheweth his handiwork."

Two rocks float through silent space, unaware that they will soon cause a scientific debate millions of miles away on Earth. On Earth, these two small asteroids added together may weigh only a few hundred pounds. The conclusions that can be drawn from them, however, might be weighty enough to overthrow a powerful theory.

That summarizes the status of a scientific debate over patterns in which meteorites fall. Evolutionary theory says that when those two asteroids smash together in space, the resulting chunks will separate during the millions of years before they ever strike Earth. As a result, meteor falls cannot possibly follow a pattern. However, a pattern has now been discovered.

Researchers from Purdue University and the State University of New York say they have discovered that 17 meteorites that struck the Earth in May between 1855 and 1895 form a broad line that extends for several thousand kilometers. Because the Earth revolves, however, the line is mathematical rather than geographical. The meteorites are classified as H chondrites. When scientists analyzed 13 of the stones, they found that each had similar amounts of rare trace elements not found in 45 other H chondrite meteorites. The other meteorites did not fall into the geographic line researchers had discovered.

The findings strongly suggest that these stones had not been drifting through space long enough to separate before they hit the Earth. If the solar system is billions of years old, there is virtually no chance of the stones remaining together. This fact suggests a young age for the solar system.

Prayer: Lord, the many wonders in the heavens do more than inspire our awe. They also bear testimony to the truth of Your Holy Word, what You have made, and our need to be restored to our Creator by grace through faith in Your innocent suffering and death for us on the cross. Amen.

Ref: Cowen, R. 1992. "Meteorites: to stream or not to stream?" *Science News*, Vol. 142, Aug. 1, p. 71.

Moon Puzzle

Psalm 148:3
"Praise ye him, sun and moon: praise him, all ye stars of light."

Our moon is moving away from the Earth at the rate of four centimeters per year. That might not seem like much. But that rate of movement away from the Earth presents problems for those who believe the Earth and moon have been around for 4.5 billion years.

At the rate the moon is receding, it would have been so close to Earth only 1.5 to 2 billion years ago that tidal friction would have melted Earth's surface rocks. Mathematical fiddlings help a little, but not enough. By mathematically increasing the rate of Earth's spin over supposed "billions" of years, and figuring in a factor for assumed different tidal rates, one can inch the Earth-moon relationship back to about 4 billion years.

Evolutionary scientists believe the problem can be solved to keep their 4.5 billion years of evolutionary history intact. But they admit that the assumptions being tried need investigation, since the matter is far from solved. In other words, evolutionists admit they have a problem making the Earth-moon relationship fit into their long-age history.

The mathematical models rule out the theory that the moon was formed billions of years ago from the same dust cloud that supposedly formed the Earth. Also ruled out is the theory that the moon was captured by the Earth's gravity. Only one explanation seems to satisfy the data – that the moon was formed relatively recently, orbiting the Earth from the time of its creation.

Prayer: I thank You, dear Father in heaven, that Your Word to us in the Bible is true and can be trusted. With the creation I will bear witness to You as Creator, and with my mouth I will tell others of salvation in Christ. In Jesus' Name. Amen.

Ref: Dye, Brad. 1988. "The moon revisited." *Creation Science Dialogue*, Spring. p. 4. Kerr, Richard A. "Where was the moon eons ago?" *Science*, Vol. 221, p. 1166.

Evolutionists Have a Back Problem

Psalm 111:10
"The fear of the LORD is the beginning of wisdom: a good understanding have all they that do his commandments: his praise endureth for ever."

Back pain is one of humanity's most common complaints. And the most common explanation one hears for back pain is that our backs hurt because we stand and walk erect.

Even high school textbooks repeat this myth. Students are taught that our backs bother us because we haven't fully adjusted to our evolution into human beings. Supposedly, when we were still getting around ape-like on all fours, our backs didn't hurt. Then we stood up. The new stresses on the spine caused it to curve forward, leading to backaches. Our backbones still have not evolved to support the stress of upright pressure. Many backache treatments are based on this assumption. So they use exercises that try to reverse the natural curvature of the back. Whether these exercises solve the long-term problem is debatable.

The Bible tells us that we did not evolve. We were created by an intelligent Creator who certainly knew how to make a backbone for human beings. Upright posture isn't our problem. Even dogs suffer back problems. That's why some therapists have been reporting good long-term results using a therapy that doesn't assume that our backs hurt because of evolution. Rather, they assume that the human spine is supposed to have a forward curve. They recommend exercises that correct the forward curvature of the spine.

Faulty assumptions that evolution is true do not provide better medicine. However, medicine is improved when it uses the accurate knowledge that human beings have been specially created by God.

Prayer: I pray, dear Lord, that You would use those of Your people who are educated in the various scientific disciplines to deliver us from the suffering and misfortunes that we endure on Earth because of the ignorance spawned by the rejection of the truth about creation. Amen.

Ref: "Your aching back: what doctors can do about it." *U.S. News & World Report,* 10/17/83, p. 85. Smail, Ronald. 1990. "The origin of back pain." *Bible-Science Newsletter,* Jan., p. 1.

Electric Hornets

Proverbs 11:2
"When pride cometh, then cometh shame: but with the lowly is wisdom."

How many hornets does it take to power a digital watch? In the case of the Oriental hornet, six hornets can generate enough electricity to run the watch.

Entomologists at Tel Aviv University have been studying Oriental hornets. They have been most interested in the outer skeleton or skin of the hornet. And they have discovered that this cuticle works as a living solar cell. When sunlight hits the hornet's cuticle, electricity is generated. To prove their point in dramatic fashion, they wired six hornets together in series. That arrangement generated enough electricity to run a digital watch for several seconds. Their research also revealed that the cuticle's electrical generation is most efficient at the temperatures in which the hornets normally operate. Different layers in the cuticle generate and store electrical current. Voltages have ranged as high as several hundred millivolts, and the current has been recorded as high as several tenths of an ampere. In effect, Oriental wasps are living semiconductors!

The researchers now find themselves in the classic situation of finding out how little they know by learning something. They report that they don't know how the hornet converts its electrical energy into a form it can use. Nor do they know how the cuticle stores the energy or transmits it.

Modern science loves to take pride in its accomplishments. However, every time science investigates God's seemingly simplest creations, scientists are reminded of their need for humility.

Prayer: Dear Lord Jesus Christ, I stand in humble awe before the works of Your hands! Help me to remember the humility that even the greatest scientists should feel in the presence of Your creation, especially when I feel intimidated by the proud pronouncements of those who claim to have proven that You are not the Creator. Amen.

Ref: Reese, K.M. 1992. "Hornet cuticle may work like organic solar cell." *Chemical & Engineering News*, Mar. 23, p. 94.

Your Invisible Eye

Proverbs 26:12
"Seest thou a man wise in his own conceit? there is more hope of a fool than of him."

Did you know that you have three eyes? In fact, someday your third eye may provide spare parts that might be necessary to repair the two eyes you normally think about.

Scientists have known for more than a century that the western fence lizard has a third eye. Its third eye can be seen as a white spot on the top of its forehead. While this third eye is an extension of the pineal gland, it has a retina, lens, and cornea. While this third eye is unable to focus light, it does sense light. Pineal eyes are also found in other lizards, frogs, and lampreys, but not in mammals. Since human beings are supposed to be evolutionarily closer to mammals, scientists did not expect to find the third eye in human beings.

When medical researchers investigating the human pineal gland compared their findings with eye researchers, they were astonished. Both the pineal and the retina make melatonin, an important chemical in our daily rhythm that also affects mood. Your pineal also makes a number of proteins that were thought to be made only by the eyes that are necessary for processing light. And like the pineal, the eye also serves as one of the body's time-keeping mechanisms. There are so many similarities between the eye and the pineal in humans that scientists hope to one day use a person's pineal as a source of genetic spare parts to treat some eye diseases!

That scientific advance might have already been made if scientists had long ago given up their belief in evolution.

Prayer: I am truly fearfully and wonderfully made, dear Lord! I thank You for the wonders You have designed and placed within me. But most of all, I thank You for Your Holy Spirit, Who has convinced me of the forgiveness of my sin by grace through faith in Your innocent suffering and death on the cross. Amen.

Ref: Julie Ann Miller. "Eye to (Third) Eye." *Science News*, Vol. 128, pp. 298-299.

Nuts to Your Health

Genesis 43:11
"And their father Israel said unto them, If it must be so now, do this; take of the best fruits of the land in your vessels, and carry down the man a present, a little balm, and a little honey, spices, and myrrh, nuts, and almonds:"

Between 70% and 90% of the calories in nuts come from fat. That's why they have traditionally been on the list of foods to avoid for those who want to avoid heart disease.

Some Christians have wondered why so many of the foods we eat seem to offer natural hazards to our health. Why would God place us in such a dietary mine field? Well, perhaps many of our fears are more the result of our ignorance than of God's design.

A recent study involving 31,200 people compared those who ate nuts at least five times a week with those who rarely ate them. To control other factors that affect heart disease, researchers used Seventh Day Adventists in their study. Seventh Day Adventists suffer fewer heart attacks than other Americans. This is thought to be due largely to their diet. When it comes to nuts, though, some eat nuts often and some rarely eat them. Researchers then kept track of who in the study group ate nuts, what kind, and how often, and who had heart attacks. They found that those who ate nuts at least once a week lowered their chance of a heart attack by 25%. Those who had nuts five times a week were half as likely to have a heart attack.

While there surely are things in the creation that can hurt us, our Creator has filled the creation with many things that benefit us as well. Modern science is showing that perhaps many modern fears about the creation arise from ignorance about our Creator's excellent designs.

Prayer: *I thank You, dear Father in heaven, for the many things You have created to help us in our lives here. Remove needless fears about the created world from my heart and mind, and increase my trust in You. In Jesus' Name. Amen.*

Ref: Raloff, Janet. 1992. "Heart risks: this is nutty." *Science News*, Vol. 142, July 25, p. 52.

The Los Lunas Rock

Deuteronomy 27:1-3a
"And Moses with the elders of Israel commanded the people, saying, Keep all the commandments which I command you this day. And it shall be on the day when ye shall cross over Jordan unto the land which the LORD thy God giveth thee, that thou shalt set thee up great stones, and plaister them with plaister: And thou shalt write upon them all the words of this law, when thou art passed over, that thou mayest go in unto the land which the LORD thy God giveth thee..."

Did someone engrave the Ten Commandments, in ancient Hebrew, on a New Mexico rock 2,000 years ago?

In 1871, Indians showed New Mexico rancher Franz Huning a basalt boulder on his land. The boulder had strange writing on it. The Indians told him that the rock, with its writing, had been there long before their tribes ever came to the area. Scholars were brought in to look at the rock. They identified the writing as paleo-Hebrew script of the style in use between 500 and 100 B.C. What did the engraving say on what has come to be known as the Los Lunas Rock? It was an engraving of the Ten Commandments. But who could have made it?

There are additional finds that are even more astonishing and seem to make the answer obvious. Above the rock is a flat mountain top. On the mountain top are ancient ruins of stone structures that seem to be designed for defense. Its design has been compared to the ruins of Lachish, in southern Judea. Another Hebrew inscription on the mountain top names the God of Israel as "our Mighty One." An astronomical petroglyph indicates a partial solar eclipse that is known to have taken place in 107 B.C. This coincides with an archaeologist's dating of the engravings to about 2,000 years ago.

Did travelers from southern Judea settle in what is now New Mexico some 2,000 years ago? Exciting evidence supports that possibility and challenges modern stereotypes about the abilities and accomplishments of the humans of 2,000 years ago.

Prayer: *I pray, dear Lord, that our proud modern age may be increasingly challenged by the evidence that shows that the human beings You have created have always been highly intelligent, curious, and capable. Amen.*

Ref: *Creation Science Fellowship Newsletter*, Aug. 1992, p. 4.

God's Hazardous Waste Experts

Job 12:7-8
"But ask now the beasts, and they shall teach thee; and the fowls of the air, and they shall tell thee: Or speak to the earth, and it shall teach thee; and the fishes of the sea shall declare unto thee."

The environment naturally produces hazardous waste without human help. Ashes from a wood fire can contain enough naturally occurring radioactivity to merit a hazardous waste classification. Uranium and other radioactive elements occur naturally in the environment and can be dissolved in ground water. Chlorofluorocarbons, manufactured for refrigeration, are also spewed from volcanoes. Also called CFCs, these chemicals are thought by some scientists to degrade the Earth's ozone layer. The fact is, nature produces more hazardous waste than does man.

Since hazardous waste is a natural part of the creation, one might expect a well-designed creation to have the ability to process hazardous materials into harmless substances. Science is beginning to discover how the Creator has designed the creation to deal with this problem.

Scientists have learned how several bacteria decontaminate water that carries dissolved uranium. One species of bacterium combines the phosphate in the water with the uranium to make uranium phosphate crystals. These crystals are stored harmlessly in the bacterium. Another species uses enzymes to make uranium ore that then settles harmlessly out of the water. Another species has been discovered that breaks down CFCs.

The creation is indeed well designed. In fact, the Creator's solutions to hazardous waste will help us learn how to clean up the messes we make.

Prayer: I thank You, dear Lord, both for the wisdom and the love with which You have designed the creation. Help modern science to learn more about Your solutions to hazardous waste so that we may take better care of the creation You have lent to us for temporary use during our lives. Amen.

Ref: "Chemistry of uranium-eating microbes." *Science News*, v. 142, Aug. 15, 1992, p. 107.

Just More Stories

2 Timothy 4:3-4
"For the time will come when they will not endure sound doctrine; but after their own lusts shall they heap to themselves teachers, having itching ears; And they shall turn away their ears from the truth, and shall be turned unto fables."

Parables are a good way to teach. Jesus often used parables to teach a lesson. Parables can even be a good way to teach science. However, parables cannot be used to establish scientific fact.

Unfortunately, Charles Darwin frequently used parables as part of his scientific method. Many modern evolutionists continue that tradition. Let's look at an example. Those who believe in evolution have tried to find an explanation for why some birds have luxurious, bright plumage. Does the peacock really need those brilliant tail feathers? Well, say evolutionists, females prefer the males with the most beautiful plumage. So, the brightest males had more babies. Evolutionists call this sexual selection.

What does the story prove? It doesn't prove anything in the scientific sense. It doesn't explain where the colors came from in the first place. It doesn't explain why females prefer showy males. Nor does it explain where peacocks originally came from. And it doesn't explain why some birds have very drab colors. The biggest problem with this explanation is that it simply begs the question. You see, many evolutionists admit that they have no good explanation for how sexual reproduction evolved in the first place.

So the evolutionists' story for how the peacock came to have such bright feathers is just a story – not scientific proof. And when it comes to stories, I prefer the stories reported in the Bible because they have our Creator's own personal guarantee of truth.

Prayer: I pray, dear Father, that You would give me a clear-thinking, critical mind informed by Your Word. Help me use these abilities to clearly see the errors of the world's thinking and better appreciate Your wisdom. In Jesus' Name. Amen.

Ref: "The Sight that Made Darwin Sick!" *Riginal View*, No. 8.

Not a Chance!

Psalm 148:7-8
"Praise the LORD from the earth, ye dragons, and all deeps: Fire, and hail; snow, and vapours; stormy wind fulfilling his word:"

Is there any such thing as chance or luck? Does anything ever happen randomly? Many people don't realize that the Bible speaks to these questions.

In late August 1992, a tornado swept through a small Wisconsin town and caused a great deal of destruction. That made it newsworthy enough. The destruction it wrought on one church in town received special notice on some national newscasts. Pictures showed the church in ruins. But the altar still stood, barely visible in the rubble. Most astonishing was the fact that the Bible still stood on its stand in its customary place on the altar. The undamaged Bible was open to where Psalm 77 reads:

"The clouds poured out water, the skies sent out a sound; your arrows also flashed about. The voice of your thunder was in the whirlwind; the lightnings lit up the world; the earth trembled and shook."

Does that sound like chance?

In Matthew 10, Jesus tells us that God is so involved in His creation that not one of the billions of sparrows in the world falls to the ground without His knowledge. Psalm 148 tells us that the entire creation praises God in everything that happens. God is so personally involved in the creation that He even instructs each wind about the speed and direction to take.

Was it simply luck that the altar was spared and the Bible was open to Psalm 77? God is involved in every detail of the creation. Let's praise God that there is no such thing as luck!

Prayer: Dear Father, I thank You that there is no such thing as luck because You are so intimately involved in the creation. Help me to cleanse my mind and speech of the pagan ideas of chance and luck. In Jesus' Name. Amen.

The Singing Lake

Psalm 66:4
"All the earth shall worship thee, and shall sing unto thee; they shall sing to thy name. Selah."

It's a quiet, cool, overcast morning at the lake. There's a hint of fog in the air and a fine, lightly falling drizzle. Then you notice the sound. It's almost a musical note, and it's coming from the lake. It's almost as if the lake were singing.

Many people report hearing a lake "sing" when there is a fine drizzle falling. Those who have never heard it are skeptical. Now science has confirmed that lakes do indeed "sing" in a fine drizzle. Scientists even know what causes it. And in this case, the scientific explanation doesn't decrease our wonder and awe at this amazing phenomenon.

Canadian scientists placed an underwater microphone at a depth of about 100 feet in a Vancouver Island lake. They placed the microphone almost 1,000 feet from shore so they would be certain not to pick up any sounds from the shore. Then they waited for the weather to change. The scientists eventually recorded the sounds of rain, hail, drizzle, and even snow hitting the surface of the lake. Yes, even snow makes a sound when it strikes the water! Their findings show that the fine drops of water in drizzle strike the surface of the lake almost as if they were tiny explosive charges. When they burst on the water, they give off a "ping." While you'd never hear a few of these droplets "ping," countless billions of them add their sound together to make the lake literally sing.

Scripture is correct in a literal sense when it tells us that everything in the creation sings praises to its Creator.

Prayer: Lord, I join the rest of creation in praising You for Your wonderful work. I praise You most of all for taking my sin upon Yourself and carrying it on the cross, into death, so that just as You rose from the dead, I might live, too. Amen.

Ref: Peterson, Ivars. "The underwater sound of rain." *Science News*, Vol. 129, p. 4.

What Can a Plant See?

Psalm 148:7a, 9
"Praise the LORD from the earth…mountains, and all hills; fruitful trees, and all cedars:"

If a plant could see, what would it look at? Amazed scientists have learned that plants can indeed see and react to their environments, just like animals!

Plants need light for photosynthesis. They grow toward the light. But this isn't what scientists are referring to when they talk about plant sight. They have discovered that plants have an additional system that allows them to react to their surroundings. Plants have pores, called stomata. Stomata allow carbon dioxide into the plant, and oxygen out, as photosynthesis takes place. They remain closed when there is little light or when water must be conserved. The more the stomata open, the faster photosynthesis takes place, and the faster the plants grow. The plant also loses water faster when the stomata are open wide.

While both blue and red light are used for photosynthesis, scientists have found that the cells that open and close the stomata respond only to blue light. The amount of blue light present turns on a pump in the cells that causes them to swell, and the stoma opens. This amazing process involves pumping protons to create electricity. In one experiment, scientists more than doubled the growth rate of orchids by providing extra blue light to open the stomata.

In the plants' ability to sense and react to their environment, we see another way that plants give glory to their Creator and show that they are not simple living things that just happened to develop.

Prayer: I thank You, Lord, that everywhere we look in the creation we see Your glory. Help me put words to this witness so that those who don't yet know the forgiveness of sins that's available through You may hear Your Word and believe. Amen.

Ref: Miller, Julie Ann. 1985. "Plant 'sight' from pores and pumps." *Science News*, Nov. 30, p. 341.

"Santa Claws" and a Tiny Dragon

Psalm 18:30
"As for God, his way is perfect; the word of the LORD is tried: He is a buckler to all those that trust in him."

The fossil record shows us examples of God's great creativity in designing living things. It also shows that life appeared suddenly on Earth in finished form. The fossil record shows us that Earth originally had a much greater variety of life. Finally, the fossil record doesn't show any evidence of creatures evolving from one type into another.

Paleontologists have been looking for fossils of unusual creatures in some of the oldest rocks that have fossils in them. In other words, these rocks in British Columbia have evidences of some of the earliest forms of life. These layers show the rich variety of life that once existed on Earth. Paleontologists have found a much greater variety of arrow worms and jellyfish than live today. But even in the earliest layers, the worms and the jellyfish are fully formed.

In addition, paleontologists have found some startling creatures. One foot-and-a-half long creature had a circular mouth with radiating teeth and claws. Another looks like a tiny, inch-long dragon. Scientists describe it as looking like the cameo of a stegosaurus. Perhaps the most unusual creature was named "Santa Claws" by one paleontologist. It has five pairs of claws attached to its head, two flaps on the side, and a tail like a beaver.

Paleontologists and Christians who believe the biblical record of creation don't dispute the facts about fossils. We object to interpretations of the fossils that needlessly contradict Scripture.

Prayer: Dear heavenly Father, even in death, brought about by man's sin, these creatures glorify You and bear witness to Your act of creation. Strengthen my faith so that I may not be intimidated by claims that contradict Your Word. In Jesus' Name. Amen.

Ref: Weisburd, Stefi. 1985. "New creatures from the Cambrian." *Science News*, Vol. 128, Nov. 16, p. 309.

False Alarm

Romans 12:4-5
"For as we have many members in one body, and all members have not the same office: So we, being many, are one body in Christ, and every one members of another."

The rain forest offers a rich variety of living things. Part of that richness can be seen in the flocks of birds that cooperate in the canopy even though the flocks are made up of several species.

The rain forest is actually two forests in one. The canopy that's formed by the tree tops supports an entirely different type of life than the part of the forest growing beneath the canopy. This lower forest is called the "understory." Flocks made up of various species of birds search for food, usually insects, in their respective territories, either in the canopy or in the understory. Each flock has its own sentinel who watches for predators and warns the rest when danger is near.

This arrangement leaves the sentinel with little time to search out its meals. So it resorts to a little trickery to feed itself. It watches the other birds forage, in addition to watching the sky. When it sees the others turn up some good insects, it shrieks the predator alarm. Then, while the other birds are distracted, it chases down the insects before the others can get to them. It appears that the other birds can tell when they are being fooled. When they hear the alarm, they immediately look at the sentinel. If the sentinel is diving for cover, they know the alarm is real. If the sentinel is diving for their meal, it's too late to do anything about it.

In effect, the other birds are trading some of their hunting skills in return for the alarm services of the sentinel. It's the Creator's wise arrangement where each creature uses its gifts for the good of all.

Prayer: Lord, help me to use the gifts and abilities You have given me in service to You and Your people. Forgive me for past selfishness, and make me more effective as Your instrument in the lives that I touch. Amen.

Ref: Miller, Julie Ann. 1986. "Tropical trickery: birds sound false alarm." *Science News*, Vol. 129, p. 40.

An Orange Elephant?

Genesis 2:19
"And out of the ground the LORD God formed every beast of the field, and every fowl of the air; and brought them unto Adam to see what he would call them: and whatsoever Adam called every living creature, that was the name thereof."

In Genesis 2:19 we read that God brought to Adam the animals He'd made to see what Adam would call them. It's been noted for generations that this task showed Adam's great intellectual abilities.

You see, Adam had to know an animal's nature to be able to give it a meaningful name. Remember that Adam had a perfect knowledge of God. As he looked at each of God's creations, he could see an expression of part of God's nature. This allowed him to identify what God had made each creature to be, so he could give a meaningful name. While this may be hard for us to understand, science now offers evidence that the names of animals and their characteristics are indeed stored together in our brains.

Researchers studied a 70-year-old woman who has had brain damage to both temporal lobes and spotty damage elsewhere in her brain. Because of the damage, she has a most unusual disability – she cannot name animals. When shown a picture of a cow, for example, or listening to its "moo," she cannot identify the animal. Nor can she remember the characteristics of animals. When asked what color elephants are, she said they are orange.

This incident provides strong evidence that the names and characteristics of animals are indeed stored together in our brains. It also should illustrate for us how Adam's abilities were far greater than ours are today. We need to treat so-called "early man" with respect.

Prayer: Father in heaven, I confess that I am so much less than You made me to be. Help me to more earnestly seek Your instruction in Your Word and order my life by Your will so that I may more clearly reflect the new life I have been given in Christ. Amen.

Ref: Bower, Bruce. 1992. "Clues to the brain's knowledge systems." *Science News*, Vol. 142, Aug. 29, p. 148.

Learning About Catastrophe

2 Peter 2:4-5
"For if God spared not the angels that sinned, but cast them down to hell, and delivered them into chains of darkness, to be reserved unto judgment; And spared not the old world, but saved Noah the eighth person, a preacher of righteousness, bringing in the flood on the world of the ungodly;"

The great Flood at the time of Noah was more than simply the world's biggest rainstorm. Scripture paints the picture of an upheaval across the face of the Earth that combined floods, landslides, volcanoes and earthquakes. As the ground literally danced with earthquakes of unimaginable intensity, hills and mountains would have flowed like pudding.

Is it possible that modern science has been blind to evidences of such upheaval? A few generations ago, geologists who believe in evolution saw no notable evidence of any great floods on Earth. Gradually they began to conclude that much of our sedimentary rock is the result of great floods.

More recently, they have started to notice evidence that mountains can literally collapse and flow like pudding. Some landslides are simple landslides where part of a mountain collapses. But sometimes the collapse turns into a flow that travels for many miles, even across flat ground. Take, for example, the Blackhawk slide at the southern edge of the Mojave Desert. Here a mass of marble fell 1.5 kilometers down and then flowed another 9 kilometers across the nearly flat desert. One description says that it looks as if the mountain simply turned to chocolate milk. Once scientists understood that this happens, they began to recognize evidence showing that this phenomenon is not unusual.

As our scientific knowledge increases, the history recorded in the Bible becomes more dramatically illustrated – not disproved!

Prayer: I pray, Lord, that as we near the time of Your return, I would be prepared, as was Noah for the flood. Help me to more intensely make my preparations, beginning today, so that I will not be caught unaware. Let my trust never waver from the forgiveness of sins that You have won for me. Amen.

Ref: Monastersky, Richard. 1992. "When mountains fall." *Science News*, Vol. 142, Aug. 29, p. 136.

Airborne Snakes

James 4:7
"Submit yourselves therefore to God. Resist the devil, and he will flee from you."

One of the things the Hawaiian Islands have been famous for is that they have no snakes. But although there are no friends or relatives for them to visit, increasing numbers of brown tree snakes are catching the next plane to Hawaii.

The snake is native to the Solomon Islands, New Guinea and Australia. Thirty years ago, Guam had no snakes either. Now there are up to 30,000 of the slightly venomous snakes per square mile. They were probably inadvertently carried to Guam by military transport planes. Guam's birds had never seen snakes before and had no natural fear of them. Since brown tree snakes are excellent climbers, most of Guam's bird species were wiped out.

The snakes' love of climbing seems to be their ticket to the Hawaiian Islands. They climb airplane landing gears and make the trip in cargo holds or wheel wells. Two snakes were recently discovered on a runway in Oahu on the same day. One died, but the other was still alive when found. Authorities are worried about more than what the presence of snakes might do to tourism. Hawaii's beautiful tropical birds would be easy pickings for the snakes, just as were the birds in Guam. Experimental snake barriers are now being devised for use at island airports.

God protects His creatures by giving them the ability to recognize their enemies. We need to recognize the disguises of our enemies, too. Our greatest enemy, Satan, has been vanquished by the Son of God, Who took the form of a man for our salvation.

Prayer: *Dear Father in heaven, I thank You that You have told us of the forgiveness of sins that has been won for us by Your Son, Jesus Christ. You have also taught us how to recognize our enemy in Your Word and taught us how to resist him. Amen.*

Ref: Yoon, Carol Kaesuk. 1992. "Hawaii's military, civilian leaders fear brown tree snake invasion." Minneapolis *Star Tribune*, July 19, p. 19A.

Mathematical Infants

Romans 12:1
"I beseech you therefore, brethren, by the mercies of God, that ye present your bodies a living sacrifice, holy, acceptable unto God, which is your reasonable service."

Does a five-month-old infant know that three is more than two? Impressive new evidence strongly supports the idea that infants have natural mathematical reasoning skills.

Researchers relied on the fact that babies tend to look longer at something new or unexpected than they do at something they have recently seen. Infants were allowed to look at either one or two Mickey Mouse dolls. Then researchers put a screen in front of the dolls. With the babies watching, they would sometimes place another doll next to the first or second doll behind the screen. Sometimes another, hidden, researcher would hide one of the dolls. Then the screen was removed.

When the screen was removed, the infant might see the same number of dolls as before, one more, or one less. Researchers found that if the same number of dolls that the infant had seen before were still present when the screen was removed, the infant quickly lost interest. However, if there was an additional doll, or one less doll, the infants spent more time looking at them. They could clearly tell that the number had changed. Other research has shown that infants can also tell when the number of beats on a drum is the same as the number of objects they are being shown.

God has built many skills and talents into us. Other research shows that infants also have language and music abilities at a surprisingly young age. That these intellectual skills are natural to us is another evidence against the evolutionary explanation for man.

Prayer: *I thank You, dear heavenly Father, for all the talents and abilities You have given me. Help me to more fully develop and use them in service to You. In Jesus' Name. Amen.*

Ref: B. Bower. "Babies Add Up Basic Arithmetic Skills." *Science News*, Vol. 142, p. 132.

An Antarctic Forest

Romans 8:38-39
"For I am persuaded, that neither death, nor life, nor angels, nor principalities, nor powers, nor things present, nor things to come, Nor height, nor depth, nor any other creature, shall be able to separate us from the love of God, which is in Christ Jesus our Lord."

As we learn more about the history of our planet, it's becoming increasingly obvious that the Earth has had many faces over its history. Remains of tropical forests have been found within 400 miles of the North Pole. Dinosaurs once lived in the tropical lushness of Alaska. At one time you could even walk to Australia from Asia.

It's also been known for some time that the great Antarctic continent was once much warmer than it is today. However, only recently have we learned that the frozen wastelands of Antarctica were forested hills not very long ago. In the 1980s, the remains of a forest were discovered along the Transarctic Mountains. The forest stretched over an area of 1,300 kilometers. When the Antarctic forest was growing, the mountainous area looked like the fjords of Chile and Norway. Scientists have continued looking for evidences of the animals that might have lived in the forest.

How long ago did the forest grow where there is now only ice and snow? Scientists have found wood from the forest. It's not very fossilized. In fact, it still floats, and it can be burned. Scientists who typically assign great ages to the Earth say that the evidence now says that the changes that took place in Antarctica happened much more rapidly than they once thought.

Our Earth is dynamic, changing much more rapidly than most ever thought. That is, unless one has believed the catastrophic history of the Earth recorded in the Bible.

Prayer: Dear Father in heaven, I thank You for reminding us that no matter what happens, You are in charge and looking out for the welfare of Your people. Keep on reminding me of this, especially when things seem out of control. In Jesus' Name. Amen.

Ref: S. Weisburd. "A Forest Grows in Antarctica." *Science News*, Vol. 129, p. 148.

Wasps Do the Biologically Impossible

Acts 17:29
"Forasmuch then as we are the offspring of God, we ought not to think that the Godhead is like unto gold, or silver, or stone, graven by art and man's device."

When it comes to wasps, you can just forget everything you learned about sexual reproduction in biology 101. There are male and female wasps. But that has very little to do with wasp reproduction.

For starters, male wasps can only reproduce female offspring. After mating, the female wasp stores the male's seed in a sac in her reproductive tract. It will only be used when the female lays her eggs on the pupa of a fly, and then only if she wants daughters. The female can control how many of her offspring will be female and how many will be male. She can squeeze the sac to fertilize an egg when it is laid. That egg will produce a female. If she decides not to fertilize the egg, it will develop into a male.

As scientists have studied this unusual system of reproduction, they have found that the story becomes even stranger. Some females produce only male offspring. Further research shows that this is due to factors inherited through the male line. The mystery is that there is no male line. Males can only produce females. Even worse, for those who believe that wasps evolved instead of being created by God, this system offers so many disadvantages to the wasp that it should never have evolved.

While the wasp is at a disadvantage in this arrangement, God has arranged things so that the wasp can survive. That survival appears to serve another of God's purposes. It challenges those who think that there is no Creator and God of the universe.

Prayer: I thank You, Lord, that You care for the universe and its creatures. I also thank You that You desire a relationship with us so much that You use what You have made as a witness to Yourself. And I thank You that You cared so much for me that You gave Your life on the cross so that my sins could be forgiven. Amen.

Ref: L. Davis. "Waspish Son-Killers and Sex-Switchers." *Science News*, Vol. 129, p. 134.

Meet Caiaphas

John 11:49-50
"And one of them, named Caiaphas, being the high priest that same year, said unto them, Ye know nothing at all, Nor consider that it is expedient for us that one man should die for the people, and that the whole nation perish not."

A remarkable discovery in Israel is shedding new light on life in Bible times. The discovery is also allowing scientists and the rest of us to get to know an important New Testament character.

Israeli archaeologists believe they may have discovered the bones of the high priest Caiaphas. If these bones do indeed belong to Caiaphas, it would be the first discovery of the remains of any major figure mentioned in the New Testament. The discovery was made accidentally in 1990 as workers were widening the road through the Peace Forest.

Researchers didn't want to release their announcement until they had satisfied themselves that such a momentous announcement was justified. The burial cave has three mentions of the name "Caiaphas." An ossuary, or bone box, within the cave was inscribed, "Joseph, son of Caiaphas." Other records identify the "Caiaphas" who condemned Jesus as Joseph, son of Caiaphas. A coin found in the cave was minted between 37 and 44 A.D. The ossuary contained the bones of six people. There were two infants, a child, a youth, an adult female, and a male about 60 years old, believed to have been Caiaphas himself.

It was from political expediency that Caiaphas said it would be better for one man to die for the people than for the entire nation to perish. He was unknowingly prophetic. Jesus Christ did die to save all of us from the eternal consequences of our sin.

Prayer: Dear Lord Jesus Christ, I thank You that You gave up heaven and endured the suffering of the cross so that my sins could be forgiven. Help me to always treasure what You have done for me and show my thanks to You in the life that I live. Amen.

Ref: Greenhut, Zvi. 1992. "Burial cave of the Caiaphas family." *Biblical Archaeology Review*, Sept./Oct, p. 29.

Wonders from Ancient Cooks

Psalm 103:1-5
"Bless the LORD, O my soul: and all that is within me, bless his holy name. Bless the LORD, O my soul, and forget not all his benefits: Who forgiveth all thine iniquities; who healeth all thy diseases; Who redeemeth thy life from destruction, who crowneth thee with lovingkindness and tender mercies; Who satisfieth thy mouth with good things; so that thy youth is renewed like the eagle's."

Many ethnic cooking traditions have their origin in health considerations. Some of those traditions are so old their origin is unknown, which strongly suggests that the ancients had a very sophisticated knowledge of food and health, some of which is only being rediscovered today.

Throughout most of history, humans have not had the luxury of refrigeration or freezing to preserve food. And even these do not provide permanent storage for food. Meats, whether cooked or raw, deteriorate even when frozen. The cold simply slows down the process. Fat spoilage in meat is a universal problem.

Chemically, fat spoilage is referred to as lipid oxidation. Lipids in meat include fat and cholesterol. Lipid oxidation does more than give meat a rancid, warmed-over flavor. Researchers believe that oxidized lipids also contribute to heart disease. Japanese research suggested that ginger, common in oriental cooking, might retard lipid oxidation in meat. Armed with this knowledge, researchers investigated whether there was any connection between the common use of ginger and Japan's very low rate of heart disease. They found that pork patties seasoned with ginger showed only one-third as much lipid oxidation as unseasoned meat.

It appears that even before so-called recorded history, people knew that ginger helped preserve meat and keep it more wholesome. This discovery is yet another refutation of the idea that ancient man was primitive and ignorant.

Prayer: I thank You, dear Father in heaven, that You have filled the creation You gave us with good things and given us the intelligence to use them. Help us to learn more of these blessings and make better use of them. In Jesus' Name. Amen.

Ref: "Season gingerly to retard rancidity." *Science News*, Vol. 129, Mar. 1, 1986, p. 137.

The Jekyll and Hyde Protozoan

Luke 18:1
"And he spake a parable unto them to this end, that men ought always to pray, and not to faint;"

Normally they bother no one. The creatures, called protozoan ciliates, quietly go about their business in water-filled tree holes, eating other microorganisms. The protozoans themselves are an important food for mosquito larvae.

As mosquito wigglers begin to populate the water, however, the protozoans turn into deadly monsters. Secretions from the mosquito larvae alert the protozoans to the presence of the larvae. They begin rapidly growing cells, and their appearance changes. The result is a completely different creature that is well-designed to infect mosquito larvae and kill them. In other words, for self-protection, the prey becomes parasite, making the would-be predator its prey. Scientists are seeking better understanding of this in the hope of developing better natural mosquito controls.

Scientists have known for some time that some microorganisms respond to predators. The most common protective response is to grow spines or some other projection. This makes the prey difficult to eat. But the strategy used by the protozoan ciliates is the most extreme self-protective response known.

There is no creature in God's creation that is so unimportant that God has failed to give it ways to make its living. Many creatures have been given what seem to be very uncreative ways to live. The fact that even these seemingly unimportant creatures have special gifts from God's creativity gives witness to His care for everything He has made.

Prayer: Dear Lord, though I cannot understand how You could have paid so much detailed attention to so many things in the six days of creation, help me to remember Your constant attention to everything when I pray. Amen.

Ref: "Revenge of the placid protozoan." *Science News*, Vol. 131, June 4, 1988, p. 363.

Truly Omnivorous

Psalm 9:10
"And they that know thy name will put their trust in thee: for thou, LORD, hast not forsaken them that seek thee."

Algae are among the world's premier plants. Scientists never expected to find an alga that feeds like a Venus flytrap. So biologists were surprised to discover that a common type of lake alga not only feeds itself through photosynthesis, but also eats bacteria.

Scientists studied the alga's feeding habits, using fluorescent latex beads about the size of bacteria. That way they could learn how many bacteria the alga would eat and what happens to the bacteria after it's eaten. They learned that an average alga will eat 36 bacteria in an hour. In other words, an alga will eat as much as one-third of its weight in bacteria every day. Scientists have also discovered two other species of alga that eat bacteria, but at only one-sixth the rate of the common lake alga. How do the immobile algae capture bacteria? The cells of the alga share a fibrous casing with flagella that extend through the top of the casing. The flagella force water into the casing, and along with it, bacteria. Once the bacterium is trapped in the casing, it is engulfed by the cell.

The ability to eat bacteria is important for the alga's existence. Like the bogs where most carnivorous plants live, clean and clear lake water is low in plant nutrients. Therefore, in each instance, the ability to eat animal life provides needed additional nutrients. In addition, bacteria are kept at safe levels for other creatures.

Unknown until recently, bacteria-eating algae are an important link in a lake's food chain. We can be thankful that our Creator is more wise than even our best scientists.

Prayer: Dear Father in heaven, I thank You that Your wisdom is above all human wisdom and that You have used Your wisdom so lavishly in the world. Help me to be less trusting of human wisdom as I learn more of the excellence of Your wisdom. In Jesus' Name. Amen.

Ref: Miller, Julie Ann. 1986. "Lake algae dine on bacteria." *Science News*, Vol. 129, Feb. 1, p. 71.

Unusual Island Love Song

1 Chronicles 16:9
"Sing unto him, sing psalms unto him, talk ye of all his wondrous works."

Tropical beauty has inspired many Hawaiian love songs. None is more unusual than the love song of Hawaiian fruit flies.

There are more than 100 species of Hawaiian Drosophila, or fruit fly. Just like everything else in that tropical paradise, these fruit flies are not the drab creatures known in mainland biology labs. Hawaiian fruit flies are larger and more brightly colored. And while mainland fruit flies do sing, Hawaiian fruit flies offer more variety in their songs and in the way they make the sounds. One of their unique songs is called the click song. Its high-pitched clicks sound something like a fingernail being dragged across a comb. Another song is made up of a complex pattern that includes pulses and trills. A third song pattern is a steady purr. A fourth pattern is made up of a low hum. Scientists describe the singing patterns as elaborate for a fly.

Scientists suspect that the fruit flies use singing as part of their mating ritual. In one encounter, a male stood behind a female with his head under her wing. He then hummed until she either accepted his advance or flew off. It may be, scientists speculate, that the female feels the male's song rather than hears it. Songs and rituals seem to vary between the species.

Scripture tells us that everything in the creation sings praise to our Creator. Often that truth is taken as figurative. As we learn more about His creatures, however, we see that the complexity of music is a gift that has been given to many things in the creation.

Prayer: I thank You, Lord, for the gift of music. Help all my words, actions, and thoughts to be a constant praise to You. Forgive me for those times I have complained, wash me clean of this worldly sin, and fill my life with thankfulness to You for salvation. Amen.

Ref: Edwards, D.D. 1990. "Unique island love songs attract flies." *Science News*, Vol. 133, p. 244.

Fossil Termites

Genesis 1:25
"And God made the beast of the earth after his kind, and cattle after their kind, and every thing that creepeth upon the earth after his kind: and God saw that it was good."

The unusual discovery of the most ancient termite nest ever found sheds light not only on termites but also on their social behavior. The nest dates back to the time of the dinosaurs.

The nest was discovered in fossilized wood from the Big Bend National Park in Texas. A paleontologist examining the wood noticed small, grain-like particles in the wood. He decided that the grains must be insect eggs. Later, other scientists examined the grains under a microscope and found that they were hexagonal in shape. That distinctive shape told them that the grains were termite droppings. Termites are the only insect that makes hexagonal droppings. And these droppings were identical to those made by modern termites. With this discovery, the holes in the fossilized wood suddenly made sense.

The wood had been tunneled out in the same way modern termites tunnel wood. The nest was in the center of the wood, just as modern termites build their nests. These ancient termites had placed their droppings around the edge of the nest. Modern termites do the same thing to plug any air leaks and prevent drafts. In short, every evidence says that termites from the time of the dinosaurs were built just like modern termites and they behaved in the same way as modern termites.

That there was no evidence of any termite evolution in this nest agrees perfectly with the Bible's claim that all things reproduce "after their kind."

Prayer: I thank You, Lord, that You have given us Your trustworthy Word. Help me to make Your Word more of an instrument in my life for Your Holy Spirit so that I may increase in faith and my life may more perfectly reflect Your will. Amen.

Ref: Weisburd, Stefi. 1986. "Oldest nest of household pest." *Science News*, Vol. 129, p. 86.

The Governor of Syria

Luke 2:1-2
"And it came to pass in those days, that there went out a decree from Caesar Augustus, that all the world should be taxed. (And this taxing was first made when Cyrenius was governor of Syria.)"

Skeptics are forever trying to find errors in the Bible so that they can discredit it as God's Word. When a faithful Bible scholar discovers some puzzle in Scripture, he never assumes that he has found an error. He assumes that he doesn't have all the information necessary to resolve the question.

As believers would expect, every so-called puzzle that has been resolved has been resolved in favor of the accuracy of the Bible. In each of these cases, those who said that they had discovered an error in the Bible were proven wrong.

One of the more interesting puzzles that was finally solved concerned Luke's account of Christ's birth. Was Quirinius really governor of Syria when Christ was born in 4 B.C.? Scholars knew that he was governor in 6 A.D. But there was no evidence that he had governed Syria in 4 B.C. Some 19th-century scholars wrote that Luke must have made a mistake with the date of the census, since Quirinius wasn't governor when Christ was born. Then, in 1912, an inscription was discovered that was dated to around 10 B.C. It said that Quirinius was governor in Syria and Cilicia around that time. In other words, Quirinius ruled the area as governor at least twice, including when Christ was born.

As God's Word, Scripture's accuracy can be trusted. Nothing has ever disproved the truth of anything in the Bible. The Bible can be trusted even when it talks about historical events. That's true even when the Bible talks about the history of the creation of the world.

Prayer: I thank You, Lord, that Your Word is completely trustworthy. I thank You that Your certain Word assures me that You fully carried the burden of my sin on the cross so that I could be forgiven. Amen.

Ref: Jackson, Wayne. 1991. "Calm confidence in the Scriptures." *Reasoning from Revelation.* Vol. 3, n. 3.

The Miracle Star

Matthew 2:1-2
"Now when Jesus was born in Bethlehem of Judea in the days of Herod the king, behold, there came wise men from the East to Jerusalem, Saying, Where is he that is born King of the Jews? for we have seen his star in the east, and are come to worship him."

What was the star the wise men followed to find the infant Jesus? About this time of year articles appear in magazines and newspapers offering many opinions. Some suggest that it was a comet. Others say it was the conjunction of two or more planets or stars. Do any of these theories offer possible explanations?

It's not a very complicated task to evaluate these explanations. All you need is a Bible and a map of the Holy Land. Matthew 2:1 tells us that the wise men came from east of the Holy Land. Verses 2 and 9 further tell us that these men had seen the star indicating His birth in the eastern sky. Despite the pictures on Christmas cards, they had not followed the star to Jerusalem. Once they arrived in Jerusalem, we learn in Matthew 2:9, the star they had seen in the eastern sky moved to lead them to the Christ Child.

Matthew clearly tells us that the wise men went from Jerusalem to Bethlehem in their search for the child. Bethlehem is due south of Jerusalem. He also tells us that the star moved over where the child was. That movement is very unlike a star. In fact, the movement of the star was even more unstar-like in that it directed them to a specific building!

These wise men from the East were accustomed to the sights of the night sky. They could tell the difference between a conjunction and a comet or a miracle star. This star was clearly created by God to announce the birth of the Savior of the world to the world He came to save!

Prayer: I know of the forgiveness of my sins, won by You for me, dear Lord, because You saw that the message of salvation was told far and wide. Increase my gratefulness for Your gift and my thanksgiving in hearing about it so that I do more to see that others hear of Your salvation, too. Amen.

Miracle Smell

Philippians 4:18
"But I have all, and abound: I am full, having received from Epaphroditus the things which were sent from you, an odour of a sweet smell, a sacrifice acceptable, well-pleasing to God."

It's commonly known that many animals are able to detect extraordinarily weak odors. The real mystery is how they can smell scents that are 1,000 times too weak to produce the chemical reactions necessary to make a scent signal.

From the chemist's point of view, our sense of smell shouldn't work as well as it does. When you smelled that wonderful dinner a few days ago, a marvel of chemical reactions was taking place in your nose. Scientists still aren't sure how we sense such a wide range of smells. It was while investigating this question that scientists may have stumbled across the answer to another question.

The receptors in our noses have to detect a certain number of scent molecules before they can trigger the chemical response that makes the signal that tells us we have smelled something. When air is drawn into your nose, an organ called Steno's duct sprays a fine mist. Scientists always thought this mist simply humidified the incoming air. Now they've discovered that the duct also makes proteins that grab onto odor molecules. Sprayed into the incoming air, the proteins collect odor molecules. Then, with their load of odor molecules, they settle onto receptors that trigger your sense of smell. As a result, even scents that are too weak to smell are concentrated by this ingenious system so that we can sense them.

Our sense of smell helps protect us, gives our food flavor, and adds richness to the experience of living. It's truly a marvel of our Creator's design.

Prayer: I thank You, dear Father in heaven, for the blessings provided though our sense of smell. Help my life to be a sweet-smelling offering to You at all times. In Jesus' Name. Amen.

Ref: Vaughan, Christopher. 1990. "Molecular odor-eaters." *Science News*, Vol. 133, p. 348.

Cyclops Ear

Isaiah 55:3
"Incline you ear, and come unto me: hear, and your soul shall live; and I will make an everlasting covenant with you; even the sure mercies of David."

All insects that hear, except one, have two ears. Some insects have ears on their legs, thorax, or abdomen. However, they all follow the same principle. Their ears are separated so that they can locate the source of the sound – except in one insect.

Scientists always thought the preying mantis was deaf. The 1,700 species of mantises have no structure that looks like an ear. Only after a long process of detailed study and testing did scientists finally discover that the mantis can hear. Further investigation finally led to the discovery of one of the most bizarre methods of hearing anywhere in the animal kingdom.

The mantis's hearing organ is difficult to call an ear. Unlike any other insect, the mantis has only one hearing organ, located in a groove underneath its thorax. The teardrop-shaped groove has a thinner cuticle than other parts of the body. Beneath the cuticle there is a relatively large air sac on each side of the groove. These sacs are connected to the insect's respiratory system. Near the top of the sac are the nerves that carry the sensation of sound to the nervous system. Scientists say that this hearing organ senses ultrasonic frequencies. When researchers played a bat-like sound to a mantis in flight, it immediately took an evasive flight path to escape the bat it thought it heard.

There is no limit to God's creativity, or to His ability to make whatever He can imagine. We should keep this in mind, especially when some human authority tells us the Bible has made a mistake.

Prayer: I marvel and praise You, dear Lord, as I consider the creation around me. Even as some would use the creation to deny Your existence, help me to more clearly see the excellence of all that You do, both materially and spiritually. Amen.

Ref: Miller, Julie Ann. 1986. "Sensory surprises in platypus, mantis." *Science News*, Vol. 129, p. 104.

The Grasshopper's Cypress

Psalm 147:7-8
"Sing unto the LORD with thanksgiving; sing praise upon the harp unto our God: Who covereth the heaven with clouds, who prepareth rain for the earth, who maketh grass to grow upon the mountains."

A grass-like plant called the grasshopper's cypress specializes in manufacturing an animal hormone. Plants make many kinds of chemicals. But the chemical production of the grasshopper's cypress is too specialized to be dismissed as accidental.

Over the years, scientists have found a few plants that make hormones that can change insect growth patterns. In these instances, the hormone affects only a limited number of insect predators.

On the other hand, the grasshopper's cypress makes the most widely occurring hormone that affects insect growth. The chemical, called a juvenoid hormone, is believed to influence insect molting and growth. When the hormone is present, a larva will molt into a larger larva. The hormone must be absent for the larva to molt into the pre-adult stage called a pupa. Even though grasshoppers don't have a pupa stage, the insect will not develop normally if it has the juvenoid hormone in its body. Scientists found that when young grasshoppers were fed grasshopper's cypress, they developed abnormally, with twisted wings and underdeveloped eggs.

The grasshopper's cypress has indeed found an effective way to control insect pests. One scientist tried to dismiss the remarkable chemistry of the grasshopper's cypress as nothing more than an accident of evolution. Biochemistry is far too complex a field to leave to chance. That's why God was personally involved in making all living things.

Prayer: Dear Father in heaven, I could never understand how You can be personally involved in everything that goes on in the creation. Yet, this is what I read in Your Word, from Genesis to Revelation. Help me to believe this comforting truth, especially when I am troubled. In Jesus' Name. Amen.

Ref: Edwards, D.D. 1990. "A leafy home for one insect hormone." *Science News*, Vol. 133, p. 326.

Letting God Create Your Day, Volume 3 Index

Page	Title
7	A Brilliant Escape
81	A Case of Bad Dates
150	A Fish Story
16	A Glowing Ballet
244	Airborne Snakes
137	A Matter of Life or Death
188	A Not-So-Hard Saying of Jesus
41	A Spider Treatment for Stroke?
161	A Star Takes a Swan Dive
158	A Sting Operation on Potato Beetles
228	A Trail That Tells a Tale
148	A Universal Graft?
213	Alaskan Dinosaurs
102	An Amazing Australian Frog
26	An Ancient Concrete Floor
56	An Ancient Cure for Malaria
159	An Ancient Industrial Secret
246	An Antarctic Forest
46	An Inside Job
242	An Orange Elephant?
169	Ancient Eye Surgery
101	Another Miracle of Sight
194	Ant Empires
152	Are We Making Progress?
47	Babylon Rent-A-Wagon
88	Baby Talk
211	Bacteria That Dissolve Steel
192	Beautiful Loyalty
204	Bee Cologne
90	Benefits from Rejecting Evolution
31	Bigger than Tyrannosaurus!
75	Biological Balance
203	Bird Barks and Lizard Growls
80	Bird of Paradise
73	Born to House Hunt
215	Brain Disappoints Evolutionary Thoughts
18	Bug Baits Bug
138	Butterfly Husbandry
44	Butterfly Physics and Stealth
104	Can Science Beat Death?
91	"Cave Men" or Men
49	Ceramic Miracles
60	Chimps Learn Math
165	Clouds of Beauty
196	Coded for a Poisonous Lifestyle

94	Coma Healing?
13	Confused Birds
140	Cooking Chemistry
160	Crunchy Medicine
142	Crystalline Silk
125	Cubic Ice
257	Cyclops Ear
224	Damascus Steel
87	Deception of the Bola
77	Deep Diving Wonders
173	Design, Not Luck
103	Destruction from Space?
132	Did Humans Evolve from Cockroaches?
64	Different Races, One Blood
65	Dinosaurs in History
122	Dissolving the Magic of Hypnosis
17	Do Spiders Feel Pain?
36	Do We Always See Clearly?
29	Do You Have "Extra" Parts?
42	Do You Have a Bad Attitude?
12	Doctor Frog
198	Dusty Stars
98	Educated Slugs
231	Electric Hornets
146	Evolutionary Medical Ethics
230	Evolutionists Have a Back Problem
134	Evolution's Influence
51	Explaining Too Much
10	Extinct Tree Is Thriving
241	False Alarm
72	Fast Rocks
100	Flipperpithicus
123	Fly Eats Toad!
84	Flying Through Water
92	Fossil Drives Evolutionists Batty
68	Fossil Hagfish Tells Story
253	Fossil Termites
218	Frog Pharmacists
74	Frozen Frogs
199	Giant Bacteria
183	Glow-in-the-Dark Flowers
97	God's Agriculture and the Stink Bug
71	God's Gift of Pets
235	God's Hazardous Waste Experts
220	God's Mind, Human Mind
200	God's Safety Valve?
149	Green Toothpaste?
76	Groceries, Ant Style

181	Hairy Drug Factories
206	Hard-Working Virus
38	Horses Before Dinosaurs
27	Hot Sharks
170	How Old Are Fossils?
195	How to Engineer a Better Turtle
45	How to Freeze a Turtle
168	Human Magnets
153	Insulated Blackbirds
121	Intelligent Animal Antics
143	Intelligent Artists
66	Is There a Third Choice?
40	Is Your Brain Really Necessary?
19	Joshua's Altar
145	Jungle Fungus Fighter
236	Just More Stories
172	Keeping Plants in the Dark
243	Learning About Catastrophe
177	Learning from Experience
202	Lighter-Than-Air Food?
105	Lightning-Like Vision
126	Magic Mirror
190	Masters of Mimicry
245	Mathematical Infants
248	Meet Caiaphas
135	Megakites
109	Millions of Dollars Per Pound
157	Miracle Bugs
256	Miracle Smell
227	Misidentification
117	Mistletoe Mimicry
116	Monkey Medicine
205	Monkey Talk
229	Moon Puzzle
144	Moth Talk
35	Multi-Legged Chemical Warfare
187	Music Grass
112	Natural Human Language
182	Natural Musical Abilities
156	New Light on Television Violence
133	No Evolution in the Galapagos Islands
237	Not a Chance!
233	Nuts to Your Health
151	Octopus School
62	Oldest Known Religious Shrine Discovered
154	One Giant Step for Mankind
82	One-Way Only
108	Pawpaw Surprise

8	Petrifying Ages
210	Physician Wasp?
163	Plants That See
164	Plastic Farms
28	"Prehistoric Man" and the Space Shuttle
54	Proof of Humans and Dinosaurs Together
55	Protective, Teaching Fathers
11	Pure Pain
120	Purpose? To Glorify God!
111	Quick Coal
219	Raccoon Bear?
39	Racing Cockroaches
186	Reptilian Housing Development Expert
48	Reverse Engineering
136	Riders on the Wind
124	Rise and Shine!
34	Robot Bugs
240	"Santa Claws" and a Tiny Dragon
217	Science Confronts Meditation Claims
21	Science Proves Teaching Abstinence Works!
222	Scientist Sea Lions
191	Scuba Gear for Bacteria
223	Secret Intelligence from the Deity
130	Seeking Rules
171	Seismic Frogs
107	Self-Esteem and Forgiveness
24	Selfishness Loses Out
179	Sepphoris
141	Smart Bacteria
129	Smart Bugs
178	Smart Lobsters
207	Smarter Than They Thought
69	Social Spiders
197	Species Confusion May Kill Ape-Man
106	Stars Are Dying Too Fast
113	Stars That Are Too Fast?
180	Stone Makers
114	Stone-Making Plants
70	Stupid Rats and Evolution
33	Sunlight on Global Warming
79	Survival of the Generous
25	Synchronized Fireflies
83	Take Heart! Be Bold!
184	Taking a Bite Out of Stress
214	Talking About the Takin
89	The Amazing Mexican Free-Tailed Bat
53	The Animal That Lives Without Air
95	The Armed and Dangerous Fungus

128	The Brooding Father Frog
174	The Deep Sea Specialist
166	The Dynamic Rain Forest
115	The First Are Made Last
175	The First Xerographer
176	The Flowering Chameleon
57	The Gardener Bowerbird
167	The Giant Helper
254	The Governor of Syria
258	The Grasshopper's Cypress
32	The Green Lacewing "Wolf"
250	The Jekyll and Hyde Protozoan
30	The Jerusalem Department of Public Works
139	The King Bee
234	The Los Lunas Rock
226	The Messiah of the Dead Sea Scrolls
162	The Miracle of Hearing
255	The Miracle Star
221	The Mot-So-Gentle Whale
185	The Mouse with Radiation Protection
50	The Panda's Thumb Revisited
118	The Parting of the Red Sea
155	The Problem of Genius
147	The Queen of All Herbs
193	The Salt of the Earth
93	The Search for Extraterrestrial Intelligence
238	The Singing Lake
208	The Spiny Postage Stamp
37	The Strange Case of the Singing Fish
225	The Talking Ear
201	The Tuatara
127	The Unique Bdellas
209	The Wandering Compass
22	The Woodpecker's Pantry
67	The World's Oldest City
96	The Wrong Pattern
212	The Young Lizard's Canteen
85	Those Dropping Science Scores
189	Tiny Superlatives and God's Love
216	Touching Love
14	Toxic Butterflies Fool Evolutionists
251	Truly Omnivorous
119	Unscientific Source?
252	Unusual Island Love Song
43	Voodoo Aspirin
61	Want to Be More Negative?
247	Wasps Do the Biological Impossible
52	Western Acupuncture

239	What Can a Plant See?
110	What Meteorologists Think About "Greenhouse"
20	What the Unborn Tells Mother
9	What's a Siphonophore?
86	Where Did Learning Come From?
58	Who Was Neandertal?
63	Why Are Human Fossils Scarce?
6	Why Gazelles Stott
23	Will Mammoths Walk the Earth Again?
59	Winged Warriors
249	Wonders from Ancient Cooks
5	Worms with Kneecaps
78	Your Brain's Produce Section
99	Your Fail Safe Heart
232	Your Invisible Eye
15	Your Portable First Aid Kit
131	Your Sixth Sense

Contact us to request:

- Biblical creation DVDs and books, including other volumes in our *Letting God Create Your Day* series.

- "Creation Moments" CDs – each containing 30 radio programs.

- Information about the many free resources available from Creation Moments.

Creation Moments
P.O. Box 839
Foley, MN 56329
1-800-422-4253
www.creationmoments.com